Design of High-Performance CMOS Voltage-Controlled Oscillators

THE KLUWER INTERNATIONAL SERIES IN ENGINEERING AND COMPUTER SCIENCE

ANALOG CIRCUITS AND SIGNAL PROCESSING

Consulting Editor*: Mohammed Ismail. *Ohio State University

Related Titles:

CMOS CIRCUIT DESIGN FOR RF SENSORS
Gudnason and Bruun
ISBN: 1-4020-7127-2
ARCHITECTURES FOR RF FREQUENCY SYNTHESIZERS
Vaucher
ISBN: 1-4020-7120-5
THE PIEZOJUNCTION EFFECT IN SILICON INTEGRATED CIRCUITS AND SENS0RS
Fruett and Meijer
ISBN: 1-4020-7053-5
CMOS CURRENT AMPLIFIERS; SPEED VERSUS NONLINEARITY
Koli and Halonen
ISBN: 1-4020-7045-4
MULTI-STANDARD CMOS WIRELESS RECEIVERS
Li and Ismail
ISBN: 1-4020-7032-2
A DESIGN AND SYNTHESIS ENVIRONMENT FOR ANALOG INTEGRATED CIRCUITS
Van der Plas, Gielen and Sansen
ISBN: 0-7923-7697-8
RF CMOS POWER AMPLIFIERS: THEORY, DESIGN AND IMPLEMENTATION
Hella and Ismail
ISBN: 0-7923-7628-5
DATA CONVERTERS FOR WIRELESS STANDARDS
C. Shi and M. Ismail
ISBN: 0-7923-7623-4
DIRECT CONVERSION RECEIVERS IN WIDE-BAND SYSTEMS
A. Parssinen
ISBN: 0-7923-7607-2
AUTOMATIC CALIBRATION OF MODULATED FREQUENCY SYNTHESIZERS
D. McMahill
ISBN: 0-7923-7589-0
MODEL ENGINEERING IN MIXED-SIGNAL CIRCUIT DESIGN
S. Huss
ISBN: 0-7923-7598-X
ANALOG DESIGN FOR CMOS VLSI SYSTEMS
F. Maloberti
ISBN: 0-7923-7550-5
CONTINUOUS-TIME SIGMA-DELTA MODULATION FOR A/D CONVERSION IN RADIO RECEIVERS L. Breems, J.H. Huijsing
ISBN: 0-7923-7492-4
DIRECT DIGITAL SYNTHESIZERS: THEORY, DESIGN AND APPLICATIONS
J. Vankka, K. Halonen
ISBN: 0-7923 7366-9
SYSTEMATIC DESIGN FOR OPTIMISATION OF PIPELINED ADCs
J. Goes, J.C. Vital, J. Franca
ISBN: 0-7923-7291-3
OPERATIONAL AMPLIFIERS: Theory and Design
J. Huijsing
ISBN: 0-7923-7284-0
HIGH-PERFORMANCE HARMONIC OSCILLATORS AND BANDGAP REFERENCES
A. van Staveren, C.J.M. Verhoeven, A.H.M. van Roermund
ISBN: 0-7923-7283-2
HIGH SPEED A/D CONVERTERS: Understanding Data Converters Through SPICE
A. Moscovici
ISBN: 0-7923-7276-X
ANALOG TEST SIGNAL GENERATION USING PERIODIC ΣΔ-ENCODED DATA STREAMS
B. Dufort, G.W. Roberts
ISBN: 0-7923-7211-5

DESIGN OF HIGH-PERFORMANCE CMOS VOLTAGE-CONTROLLED OSCILLATORS

LIANG DAI
Prominent Communications, Inc.
San Diego, California, USA

RAMESH HARJANI
Department of Electrical and Computer Engineering
University of Minnesota
Minneapolis, Minnesota, USA

Springer Science+Business Media, LLC

Library of Congress Cataloging-in-Publication Data
Liang Dai & Ramesh Harjani
Design of High-Performance CMOS Voltage-Controlled Oscillators
ISBN 978-1-4613-5414-7 ISBN 978-1-4615-1145-8 (eBook)
DOI 10.1007/978-1-4615-1145-8

Printed on acid-free paper.

Contents

List of Figures ix
List of Tables xv
Preface xvii
Acknowledgments xix

1. INTRODUCTION 1

2. INTRODUCTION TO PLLS 9
   1 Introduction 9
   2 PLL Basics 10
   3 A Linear Model for PLLs 18
   4 Conclusions 26

3. PHASE NOISE AND TIMING JITTER 27
   1 Phase Noise 27
   2 Timing Jitter 30
   3 Phase Noise vs. Timing Jitter 33
   4 Conclusions 37

4. REVIEW OF EXISTING VCO PHASE NOISE MODELS 39
   1 Challenges in Oscillator Phase Noise Analysis 39
   2 Leeson's Model 40
   3 Razavi's Model 46
   4 Hajimiri's Model 49

5. UNIVERSAL MODEL FOR RING OSCILLATOR PHASE NOISE 55
   1 Comparison and Analysis of Ring Oscillator Phase Noise 55

1.1 VCO Circuit Diagrams 56
1.2 Phase Noise Analysis 56
1.3 Measurement Results 61
2 Modified Linear Model 63
2.1 Theoretical Analysis 64
2.2 Simulation Results 68
2.3 Measurement Results 69
3 Q-factor for Ring Oscillators 71
4 Noise Up-Conversion 74
4.1 Theoretical Analysis 75
4.2 Simulation Results 78
4.3 Measurement Results 80
5 Power Supply / Substrate Noise 81
5.1 Theoretical Analysis 81
5.2 Simulation Results 83
6 Conclusions 85

6. NEW RING VCO DESIGN 87
1 Introduction 87
2 Phase Noise Overview 87
3 Circuit Design 89
4 Analysis of Circuits with Hysteresis 92
5 Simulation and Measurement 100
6 Conclusions 106

7. PLL DESIGN EXAMPLES 107
1 PLL with Ring VCO 108
1.1 Ring VCO 108
1.2 Charge Pump and Loop Filter 109
1.3 Phase/Frequency Detector 119
1.4 Prescaler and Frequency Divider 120
2 LC VCO 126
3 Simulation Results 136
4 Measurement Results 142
5 Conclusions 145

8. CONCLUSIONS 149
1 Research Contributions 149

2 Summary 150

List of Figures

1.1 Weaver image reject receiver 2
1.2 A simplified block diagram for clock/data recovery circuits 4
1.3 Cross talk due to phase noise for wireless receivers 5
1.4 A parallel resonator 6
1.5 A ring oscillator 6
2.1 Illustration of a PLL with $N = 3$ 9
2.2 A simplified block diagram for a typical PLL 10
2.3 Characteristics of a multiplier as a PD 11
2.4 Characteristics of an XOR gate as a PD 12
2.5 A phase/frequency detector (PFD) 13
2.6 A charge pump PLL 14
2.7 Example inputs and outputs for PFD (a) $f_R < f_V$ (b) V is lagging R 14
2.8 Illustration of dead zone 15
2.9 PD characteristics with dead zone 16
2.10 Elimination of dead zone (a) with zero phase error (b) with finite phase error 17
2.11 Periodic output current from the charge pump due to current mismatch 17
2.12 A linear model for a PLL in locked state 18
2.13 A linear model for a charge pump PLL with noise sources 20
2.14 A PLL open-loop transfer characteristic with and without lead-lag compensation 21
2.15 Comparison of closed-loop and open-loop response for a charge pump PLL 22

2.16	Derivation of noise transfer characteristic for VCO	23
3.1	Illustration of phase noise and timing jitter	28
3.2	Definition of SSB phase noise	29
3.3	A typical spectrum for SSB phase noise	29
3.4	Jitter measurement (a) self-referred (b) input-referred	31
3.5	Self-referred jitter of a stand-alone VCO	32
3.6	Self-referred jitter vs. time interval for a VCO	32
3.7	Self-referred jitter vs. time interval for a PLL	33
4.1	A linear feedback system	39
4.2	RLC parallel network	41
4.3	Magnitude of the impedance across the RLC tank	42
4.4	Model for an LC tank with an active device $-R$	43
4.5	Additive noise to phase noise conversion	45
4.6	PSD plots for input device noise and phase noise	46
4.7	Barkhausen criteria	47
4.8	Waveforms in a 3-stage ring oscillator	49
4.9	Perturbation of voltage waveform due to a current impulse at (a) the zero crossing (b) peak	50
5.1	Differential inverter with Maneatis loads	57
5.2	Differential inverter with single PMOS loads	57
5.3	Coupled ring oscillator (a) inverter stage (b) strong and weak inverters (c) weighted current adder (d) $\overline{V_{cntrl}}$ generator	58
5.4	Current-starved inverter oscillator	58
5.5	Internal differential voltage waveforms for ring oscillators with Maneatis load, saturation load and triode load	60
5.6	Die photograph of the test chip	62
5.7	SSB phase noise of the ring oscillators	63
5.8	Modified linear model for a 3-stage ring VCO	64
5.9	Sinusoidal waveform clipped by power supplies	65
5.10	Waveform with soft slipping	66
5.11	SSB phase noise vs. V_{pp}	68
5.12	The bias and a delay cell for a ring oscillator with a source coupled pair and symmetric loads	70
5.13	A three-stage coupled ring oscillator	70
5.14	Chip microphotographs for (a)Maneatis ring oscillator (b)coupled ring oscillator	71

5.15	SSB phase noise comparison between the Maneatis oscillator and the coupled ring oscillator	72
5.16	Effective Q factor as a function of the f_T of the process	74
5.17	Bias structure for an N-stage ring oscillator	75
5.18	Up-conversion mechanism for low frequency noise in the tail and bias	76
5.19	The bias and a delay cell for a three-stage ring oscillator	79
5.20	Simulation of frequency variation vs. bias current variation	79
5.21	Phase noise with different by-pass capacitors	80
5.22	A VCO block diagram (a) traditional view (b) a rigorous view	81
5.23	A three-stage differential ring oscillator	82
5.24	Waveforms showing the six equivalent states for a three stage oscillator	82
5.25	Sideband PSD when sinusoidal ripple is added to the power supply	84
6.1	Illustration of major noise sources in a traditional fully differential ring oscillator	88
6.2	VCO schematic (a) top-level (b) logic-level for delay cells (c) transistor-level for delay cells	90
6.3	Channel resistance of an NMOS transistor with $V_{ds} = 0$ vs. gate voltage.	91
6.4	A simplified delay cell without delay control	92
6.5	An example case for a step response	93
6.6	The threshold of the delay cell as a function of β_2/β_1	94
6.7	Simulated hysteresis curve for the delay cell	94
6.8	An N-stage ring oscillator with hysteresis delay cells	95
6.9	Waveforms at nodes a and b in Figure 6.8	95
6.10	$T/(RC)$ as a function of V_{th}/V_{dd} for a 2-stage ring oscillator	97
6.11	Sensitivity of period with respect to threshold voltage for a 2-stage ring oscillator	98
6.12	$T/(RC)$ as a function of V_{th}/V_{dd} for a 3-stage ring oscillator	99
6.13	Sensitivity of period with respect to threshold voltage for a 3-stage ring oscillator	99
6.14	Waveforms for ring oscillators with different number of stages	100
6.15	Quadrature output waveforms	100
6.16	Die photo	101
6.17	Circuit board for VCO testing	101
6.18	Measured control characteristic for the ring VCO	102

6.19 Test setup for common mode rejection at the control terminals (a) common-mode interference (b) single-ended interference 103

6.20 Spectrum with interference at the control terminals 103

6.21 Sideband amplitude vs. frequency of interference 104

6.22 Supply and substrate noise rejection by differential pairs (a) ideal case (b) with parasitic capacitance 104

6.23 Sideband amplitude vs. frequency for supply noise 105

6.24 Measured single-sideband phase noise 106

7.1 Ring oscillator 108

7.2 Frequency control characteristic for the ring oscillator 109

7.3 A linear model for our charge pump PLL 110

7.4 A cross-section view of an on-chip capacitor with sandwiched layers 111

7.5 Open-loop and Closed-loop transfer functions for the PLL with a ring VCO 112

7.6 Frequency step response for the charge pump PLL 113

7.7 Differential charge pump and loop filter 113

7.8 Charge pump model 114

7.9 Amplifier for CMFB 116

7.10 Folded cascode opamp with CMFB and source followers 117

7.11 A single-ended active loop filter 118

7.12 Loop filter noise model at 100kHz 119

7.13 PFD with no dead zone 120

7.14 Circuit to generate complementary signals 120

7.15 Pulse swallow frequency divider 122

7.16 Top level diagram for feedback divider 122

7.17 Divide-by-2 circuit 123

7.18 Divide-by-600 circuit 124

7.19 A divide-by-2 toggle flip-flop 124

7.20 A D-type flip-flop 125

7.21 Example waveforms for the 2-input *OR* gate 126

7.22 Schematic for the 7-input *OR* gate 126

7.23 Schematic for the LC VCO 127

7.24 Cross-section view of an accumulation mode MOS varactor 127

7.25 C-V characteristic for an accumulation mode NMOS varactor 128

7.26 A cross-section view of varactors with shared bottom plates 128

7.27 A cross-section view of varactors with shared top plates 129
7.28 A model for an RLC tank with parasitic capacitance between the N-well and substrate 129
7.29 Real and imaginary parts of $Z(j\omega)$ for different R values 130
7.30 Layout for a symmetrical spiral inductor 131
7.31 A model for an on-chip spiral inductor in a CMOS process 132
7.32 Generation of eddy current 134
7.33 Simulated signals at PFD inputs: reference and feedback divider output 137
7.34 Simulated *UP* and *DOWN* voltages 137
7.35 Simulated control voltages 138
7.36 Simulated control voltages with respect to *UP* and *DOWN* signals 139
7.37 Simulated complimentary outputs of the ring VCO 139
7.38 Comparison between an accumulation mode varactor and an NMOS transistor 140
7.39 Simulated complimentary outputs of the LC VCO 141
7.40 Simulated frequency characteristic for the LC VCO 141
7.41 Simulated complimentary outputs of the LC VCO 142
7.42 A micro photo of the PLL test chip 143
7.43 Self-referred jitter of the output clock 144
7.44 Self-referred jitter of the reference clock 145
7.45 VCO output waveform at 1.92GHz 146
7.46 Spectrum at the feedback divider output 147

List of Tables

1.1	SONET optical line rates	3
5.1	Calculated Q_{eff} for processes available from MOSIS	74
5.2	$\Gamma_{RMS,eff,thermal}$ and $\overline{\Gamma_{eff,flicker}}$	78
7.1	Device sizes for the circuit in Figure 7.1	108
7.2	Device sizes for the charge pump	114
7.3	Device sizes for the amplifier used in the CMFB	116
7.4	Device sizes for the folded cascode opamp	117
7.5	Device sizes for the complementary signal generator	121
7.6	Device sizes for prescaler	123

Preface

Phase-locked loops (PLL) are used extensively in communications systems, such as frequency synthesizers and clock recovery circuits. Voltage-controlled oscillators (VCO) are important building blocks in PLLs. The random fluctuations in the output phase of the oscillator, in terms of jitter or phase noise, are extremely undesirable in most applications. The VCO is the major contributor to the PLL phase noise at frequencies outside its loop bandwidth.

This book presents a phase noise modelling framework for CMOS ring oscillators. The analysis considers both linear and nonlinear operations. It indicates that fast rail-to-rail switching has to be achieved to minimize phase noise. Also the flicker noise in the bias circuit can potentially dominate the phase noise at low offset frequencies. Therefore, for narrow bandwidth PLLs, noise up conversion from the bias circuits should be minimized. We define the effective Q factor (Q_{eff}) for ring oscillators and predict its increase for CMOS processes with smaller feature sizes. Our phase noise analysis is validated via simulation and measurement results.

Digital switching noise coupled through the power supply and substrate are the dominant source of clock jitter. Improving the supply and substrate noise immunity of a PLL is a challenging job in hostile environments such as a microprocessor chip where millions of digital gates are present. This book analyzes the impact of the supply and substrate noise on the oscillator phase noise, and suggests that increasing the signal swing is also an effective way of reducing the jitter due to the supply and substrate noise.

We have also analyzed ring oscillators with memory delay cells. It is shown that the amount of jitter can be reduced by introducing memory into the delay cells. In particular, we have calculated the noise sensitivity for 2-stage and 3-stage ring oscillators with memory delay cells.

We have designed a two-stage ring VCO in a $0.35\mu m$ CMOS technology with differential control and quadrature outputs. It minimizes phase noise by using positive feedback in the delay cells to increase the transition speed. It incorporates a differential frequency control topology to reduce the impact of common-mode noise coupled through the power supply and the substrate. In particular, the use of differential frequency control in combination with a differential charge pump and a differential loop filter improves the phase noise performance of PLLs substantially. Additionally, it eliminates the cumbersome differential-to-single-ended converter.A test chip with this circuit operates with a power supply voltage of 3.3V. The measured common-mode noise rejection at 1MHz is 32dB better than for a single-ended control topology. The measured phase noise is -117dBc/Hz at a 1MHz offset from the 973MHz center frequency. This phase noise number is comparable to that of LC tank oscillators with on-chip spiral inductors, clearly indicating that low-phase-noise ring-type VCO's can be designed and manufactured. We have also designed a complete PLL with a ring oscillator based on the above design. It is fabricated in a $0.25\mu m$ CMOS technology. Additional measurement results are presented throughout this book.

LIANG DAI AND RAMESH HARJANI

Acknowledgments

We would like to thank Nat Dwarakanath and his group members at Lucent Technologies, Allentown, PA for their support for this work. In particular, Will Wilson, who is a Distinguished Member of Technical Staff at Lucent Technologies, shared his knowledge and made great contributions to our research through helpful discussions.

Chapter 1

INTRODUCTION

Phase-locked loops (PLL) are widely used in the communications systems. The first PLL was reported as early as in 1932 [de Bellescize, 1932]. Ever since then PLLs have been used for carrier recovery for televisions and FM demodulation for stereo radios. With the development of integrated circuits (IC), low cost and high quality PLLs have become available and they continue to play an important roll in today's communications devices.

As the wireless communications market has boomed in recent years, PLLs have found their application for frequency synthesis in wireless transceivers. With the help of an accurate frequency reference, a PLL can generate carriers at precise frequencies. In contract to the traditional frequency tuning techniques available on old radios, PLL-based frequency synthesizers can track the exact frequency digitally, precisely and quickly.

An example wireless receiver architecture is shown in Figure 1.1. This architecture is known as the *Weaver image reject receiver*, named after his inventer, D. K. Weaver in 1956 [Weaver, 1956]. There are two frequency synthesizers in this architecture, providing two LO (Local Oscillator) frequencies at ω_1 and ω_2. Quadrature outputs are needed for both of these LO's, namely an in-phase *(I)* clock and a quadrature-phase *(Q)* clock. There is a phase difference of 90° between the I and the Q channels. The incoming radio frequency (RF) signal at ω_{RF} is first mixed with the first LO at ω_1 and later mixed with the second LO at ω_2. The resulting intermediate frequency (IF) signal, being either at baseband or any non-zero frequency, is a fixed frequency and has to be carefully planned for signal amplification and channel selection. Therefore, the LO frequencies ω_1 and ω_2 have to be well controlled so that the RF signal in the desired channel is converted to IF instead of its adjacent channels or any other neighboring signal channels.

Similar requirements have to be met for transmitters. The baseband signal modulates the carrier which is generated by the frequency synthesizers. In order to transmit the signal in the desired channel, the carrier frequency has to be precisely controlled with the synthesizer.

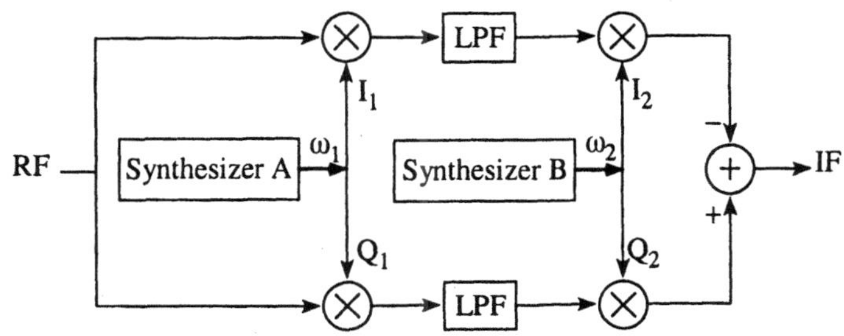

Figure 1.1. Weaver image reject receiver

Modern communications protocols allocate closely located channels at very high frequencies. Let us consider Bluetooth, a short range wireless protocol, as an example. There are 79 channels allocated from 2.402 GHz to 2.480 GHz and each channel is only 1 MHz away from its neighbors. Therefore, the phase noise of the LO has to be low enough so that it does not spill its energy over to the next channels. Otherwise, crosstalk between adjacent channels will occur. If a standalone oscillator is used to generate the LO, a Q factor of at least 10^4 is needed. This can be achieved only by a crystal oscillator. However, Bluetooth requires that the LO frequency has to be precisely tuned over a 79 MHz band. This is far beyond the frequency tuning range of any crystal oscillator. Additionally crystals do not have a resonant frequency as high as 2.4 GHz. Therefore, the LO frequency can only be generated with the help of phase-locking techniques.

PLLs are negative feedback systems whose output frequency can be digitally controlled with the help of a precise clock at its input as a reference. Its phase noise is suppressed within the loop bandwidth by negative feedback. Therefore, they are ideal devices for frequency synthesis.

PLLs have also found applications in data communications as clock recovery circuits. At the receiver end of the channel, the digital data have to be sampled at the same rate as they are sent at the transmitter end. Therefore, the clock has to be recovered from the data sequence with the help of a PLL.

Let us take a look at SONET (Synchronous Optical NETwork), a standard in data communications where PLLs play a crucial role. SONET defines optical carrier (OC) levels and electrically equivalent synchronous transport signals (STSs) for fiber-optic based transmission hierarchy. The standard SONET line rates and STS-equivalent formats are shown in Table 1.1.

Table 1.1. SONET optical line rates

Optical Carrier Level	Electrical Equivalent	Line Rate (Mbps)
OC-1	STS-1	51.84
OC-3	STS-3	155.52
OC-12	STS-12	622.08
OC-24	STS-24	1244.16
OC-48	STS-48	2488.32
OC-192	STS-192	9953.28
OC-768	STS-768	39813.12

In SONET the average frequency of all clocks in the system will be the same or nearly the same. Every clock can be traced back to a highly stable reference signal. Thus, the STS-1 rate remains at a nominal $51.84 Mbps$, allowing many synchronous STS-1 signals to be stacked together when multiplexed without any bit-stuffing. Thus, the STS-1's are easily accessed at a higher STS-n rate.

As a synchronous network, SONET is organized with a master-slave relationship with clocks of the higher level nodes feeding timing signals to clocks of the lower level nodes. SONET eliminates the need for sync characters and message framing for clock synchronization between equipments. Instead, SONET relies on a synchronization network that transports timing references between locations through the signals that the network elements (NE) at a higher level send to the network elements at a lower level. By this synchronization hierarchy, each NE can be traced up to a primary reference source, a Stratum 1 atomic clock with extremely high stability and accuracy. Therefore, precise timing recovery is critical in SONET.

After being detected by a photo detector and amplified by a pre-amplifier, the incoming data is processed by a clock/data recovery circuit (CDR) as shown in Figure 1.2. The clock information is first recovered by the clock recovery circuit, and then it is used to sample the incoming data stream. During this process, the jitter in the input data is suppressed. Meanwhile, some jitter is generated by the CDR due to the circuit noise. SONET has stringent requirement on jitter transfer, jitter generation and jitter tolerance. This becomes a challenging task for high-speed communications. OC-48, for example, has a data rate of about 2.5 Gbps as listed in Table 1.1. The clock period is as short as 400 ps. As the operating speed of the CDR is pushing towards the speed limits of current CMOS technology, the rise and fall times take a considerable portion of the clock period. Any slight phase variation in the sampling clock may result in a wrong decision.

In all these applications, the phase noise or the timing jitter at the PLL output severely degrades the system performance. It causes cross talk in wireless communications and increases bit error rate (BER) in data communications.

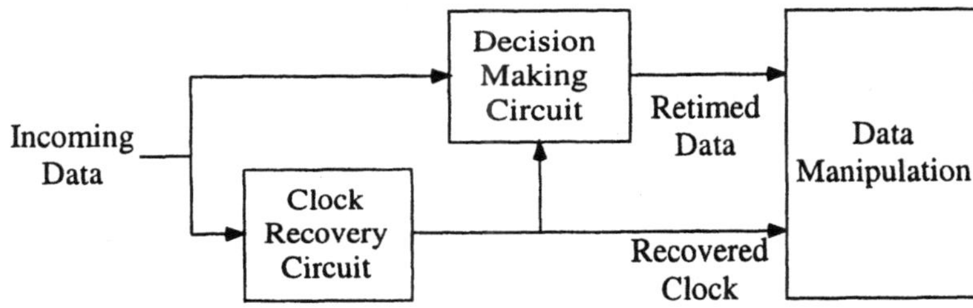

Figure 1.2. A simplified block diagram for clock/data recovery circuits

The problem becomes worse in high frequency applications. An illustration of the cross talk due to phase noise for wireless receivers is shown in Figure 1.3. Without losing generality, we consider the case for a homodyne receiver where the RF signal is directly down converted to baseband. The same argument is valid for heterodyne receivers where the RF is converted to a non-zero IF frequency. Without phase noise, the signal in the desired channel is down converted to frequencies around DC, while all the out-of-channel interferers are eliminated by the channel selection filter. However when phase noise is present in the LO, it will cause the bandwidths of both the desired signal and the interferer to become wider. Hence, some of the energy from the other channels may expand into the baseband and overlap with the desired signal spectrum. The signal will be corrupted due to the spectral overlap and can no longer be recovered by the channel selection filter. This results in cross talk between channels. In many cases, the power of the interferer can be much larger than that of the signal in the desired channel. Therefore, the cross talk due to the LO phase noise has to be minimized.

Due to its high level of integration, CMOS technology is the dominant technology for the digital circuits that process the baseband signal. Therefore it is desirable to implement the RF or high-speed analog front-end in CMOS too so that a single-chip solution is possible for systems-on-a-chip designs. This implies that PLLs that are designed with digital CMOS processes are highly desirable.

A typical PLL includes a phase/frequency detector (PFD), a charge pump, a low-pass loop filter, a voltage-controlled oscillator (VCO) and a frequency divider. The VCO is the major phase noise contributor outside the PLL loop bandwidth. Current digital CMOS technologies do not provide very high-Q on-chip inductors. Hence the VCO phase noise is poor and usually is the limiting factor for many applications where low phase noise is required. Furthermore, accurate modelling of spiral inductors is still not an easy task. Even though the process variation of the inductance is small, the prediction of the inductance

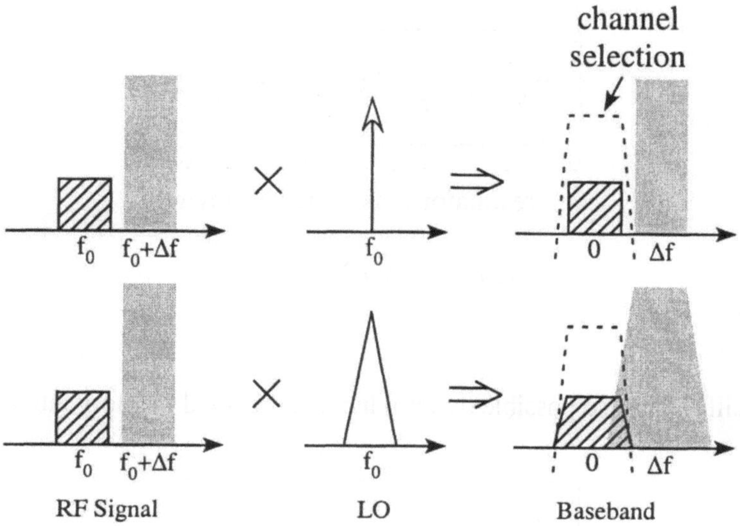

Figure 1.3. Cross talk due to phase noise for wireless receivers

value may be off without a few fabrication runs. Therefore, the VCO may fail to cover the desired frequency band in the first try.

There are two types of VCOs that are promising candidates for CMOS PLL solutions. One is the resonator-based VCOs which include on-chip inductors and varactors. The other type is resonatorless VCOs. A resonator-based VCO has a resonator tank as a frequency selective element consisting of inductors and capacitors. The energy loss in the tank has to be compensated for by active devices. An example of a parallel resonator is modelled in Figure 1.4. The resistor R models the resistive loss in the tank, and the negative resistor models the active device which provides energy into the tank. Once the energy loss is equal to the energy provided by the active device, a stable oscillation can be sustained.

A popular type of resonatorless VCO is ring oscillators. As shown in Figure 1.5, a ring oscillator is formed by connecting odd number of inverters as a ring. Due to the inversion of each stage, the odd number of inverters have a phase shift of 180° in total. Additionally, each inverter is a low-pass filter with some intrinsic delay. The additional delay contributed by all the stages add up to 180° at a certain frequency ω_0. Therefore, positive feedback is formed at ω_0, and oscillation results. If fully differential stages are used, even number of stages can also form a positive feedback loop by flipping the two input connections at one of the stages. In general, the minimum number of stages for a ring oscillator is 3. However, as we will discuss in this book, a two-stage

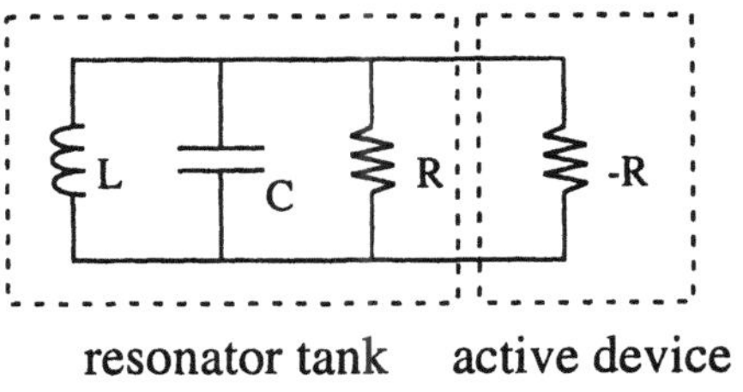

Figure 1.4. A parallel resonator

ring oscillator is also possible by introducing additional memory into the delay cells.

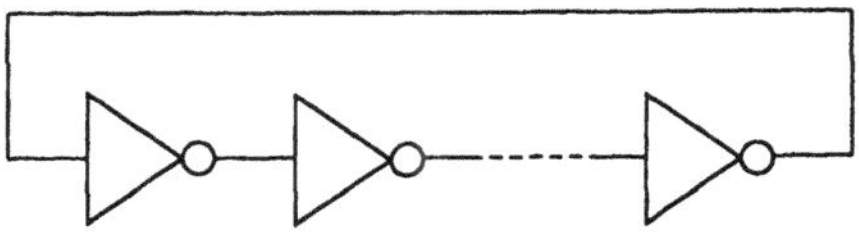

Figure 1.5. A ring oscillator

Resonatorless VCOs are fully compatible with the standard digital CMOS technologies. They occupy small chip area and have a large frequency tuning range. However, due to the lack of a high-Q resonator, their phase noise performance is poor. On the other hand, the resonator-based VCOs have improved phase noise with the help of an on-chip inductor. Even though the Q factor of an on-chip spiral inductor is still low compared to that of its discrete component counterparts, its existence helps to extend the VCO into some of the applications where low phase noise is required. Unfortunately they take much larger chip area and have a small frequency tuning range.

This book is organized as follows. In Chapter 2 we first introduce some basic knowledge about PLLs. We briefly discuss each of the building blocks in a PLL and apply linear analysis for phase-locked state. Based on linear analysis we try to understand how each building block contributes phase noise to the PLL. We will also discuss the relationship between phase noise and timing jitter. In Chapter 3, we introduce definitions for single-sideband (SSB) phase noise and timing jitter, and discuss their relationship. Then, in Chapter 4, we review some of the existing phase noise models for oscillators. By discussing the advantages and disadvantages of these models we introduce our motivation for this work. In Chapter 5 we develop a theoretical frame work for CMOS ring oscillator phase noise analysis. Our VCO phase noise analysis method serves as a good compromise between accuracy and computation complexity. Later we apply our theoretical analysis and present a new CMOS ring oscillator design in Chapter 6. Additionally we develop jitter analysis to include ring oscillators with memory delay cells. Two PLL designs are presented in Chapter 7 with measurement results . Finally we provide some conclusions in Chapter 8.

Chapter 2

INTRODUCTION TO PLLS

1. Introduction

The function of a PLL is to generate an output clock whose phase is locked to that of the input reference clock. In order to satisfy this condition, the output clock frequency either has to be equal to the input clock frequency or has to be a multiple of the input clock frequency, i.e. $f_o = Nf_i$ where f_o is the output frequency and f_i is the input frequency. This is illustrated in Figure 2.1 for an example of $N = 3$. A PLL is said to be *locked* when its output frequency is equal to a multiple of its input frequency and the phase error is kept to some finite and small value. In other words, both the frequency and the phase of the output clock have to be synchronized with the input clock in the locked state.

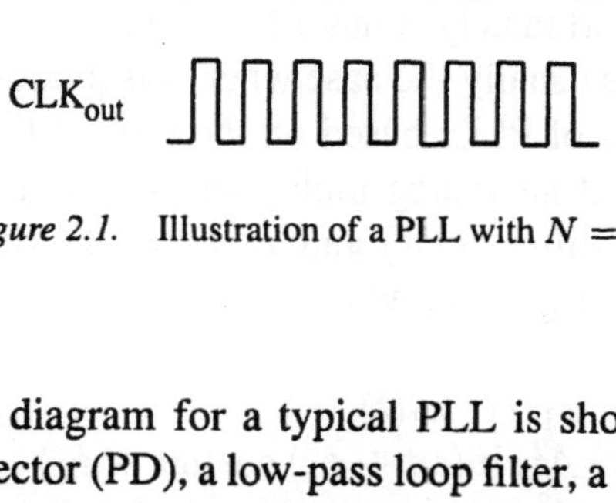

Figure 2.1. Illustration of a PLL with $N = 3$

A simplified block diagram for a typical PLL is shown in Figure 2.2. It consists of a phase detector (PD), a low-pass loop filter, a VCO and a frequency divider. A phase detector is a device whose output voltage is proportional to the phase difference between its two inputs. More often a phase/frequency detector (PFD) is used instead in order to reduce the lock capture time. This will be discussed in more details in later sections. For high-frequency PLLs, the first few stages of the frequency divider may experience the greatest difficulty in handling the high-frequency output from the VCO. Therefore, more "analog-like" circuits are often used for the input stages of the divider instead of the CMOS logic cells, and this functional block is usually called a *prescaler*.

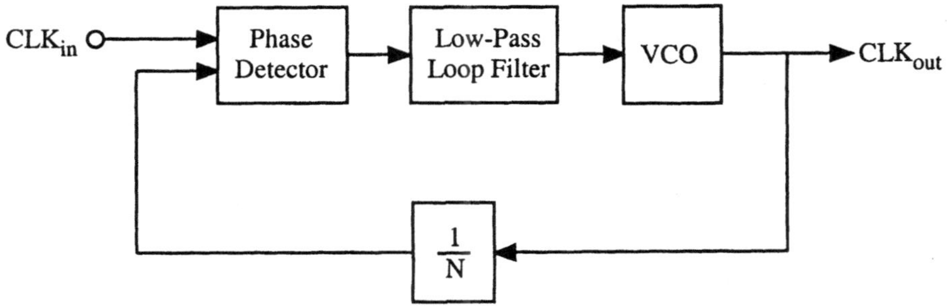

Figure 2.2. A simplified block diagram for a typical PLL

In this chapter, we will discuss each building block for a PLL briefly in Section 2. PLLs are nonlinear systems in general. However, they can be approximated by a linear model in the phase-locked state. Therefore, we review linear analysis of PLLs in Section 3. The linear model helps understand the noise transfer from each component to the output of the PLL. This is very useful in low-phase-noise PLL designs.

2. PLL Basics

A PLL is a highly nonlinear system in general, since the PD has a nonlinear characteristic over a large phase input range. As the phase error increases, it will wrap around at some point. Hence, all PDs have a period characteristic for the input phase. Fortunately it has a fairly linear characteristic for a small phase input, and this is usually the case when it is phase-locked.

There are two types of PLLs based on the type of the PD used. A PLL is called an analog PLL if an analog multiplier is used as a linear PD. For two inputs of $x_1(t) = A\sin(\omega t + \phi_1)$ and $x_2(t) = A\cos(\omega t + \phi_2)$, the analog multiplier output is their product which is given by

$$\begin{aligned} y(t) &= x_1(t) \cdot x_2(t) \\ &= A^2 \sin(\omega t + \phi_1)\cos(\omega t + \phi_2) \\ &= \tfrac{A^2}{2}\left[\sin(2\omega t + \phi_1 + \phi_2) + \sin(\phi_1 - \phi_2)\right] \end{aligned} \tag{2.1}$$

The first term in Equation 2.1 is the high frequency component and it is usually filtered out by the low-pass loop filter. Hence only the second term is left, and Equation 2.1 can be approximated by Equation 2.2 for small phase errors. Because of the linear characteristic for small phase error, a multiplier can be used as a PD, and the corresponding PD gain is given by Equation 2.3.

$$y(t) \approx \frac{A^2}{2}\sin(\phi_1 - \phi_2) \approx \frac{A^2}{2}(\phi_1 - \phi_2) \tag{2.2}$$

$$K_{PD} = \frac{A^2}{2} \tag{2.3}$$

The characteristic of a multiplier as a PD is shown in Figure 2.3. Here we assume that the phase of a sine wave is compared with the phase of a cosine wave. Therefore, a systematic phase error of $\pi/2$ exists. It is seen from the figure that it is monotonic only for input phase error between $\pm\pi/2$, and the linear approximation is only valid for small phase inputs. For large inputs, it is not only nonlinear, but also non-monotonic.

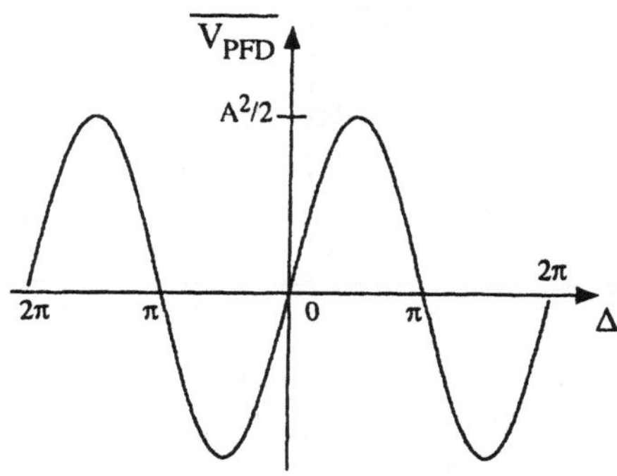

Figure 2.3. Characteristics of a multiplier as a PD

It is also seen from Equation 2.3 that the PD gain is related to the signal amplitude A. In the case where the signal amplitudes are not well controlled, the PD may experience large variation of its gain, and this may result in undesirable loop characteristics.

Another problem may occur during the phase lock capture. Before the output clock is locked to the right frequency, the PD sees two different frequencies at its inputs, namely ω_i and ω_o. By the mixing operation, the PD generates two frequencies at its output, $\omega_i + \omega_o$ and $|\omega_i - \omega_o|$. The average of the PD output at both frequencies are 0. This means that it does not provide any frequency information that can help pull the VCO towards the right frequency to help lock. Therefore, the capture process may take a long time or even fail.

A PLL is called a digital PLL when logic gates are used. The name is somewhat misleading, since the PLL is still not a complete digital system, since the VCO, the loop filter and even the prescaler are still analog components.

An XOR gate can be used as a PD, since it is nothing but a digital form of a multiplier. Now that signal amplitudes are fixed at 0 and V_{dd}, the PD gain is no longer dependent on the signal amplitude. However, a systematic phase error of $\pi/2$ still exists, and it does not provide any frequency information either during the phase capture process. Figure 2.4 shows the characteristic of an XOR gate used as a PD. Because of the low-pass loop filter, only the average of the output voltage will be used to control the VCO frequency. It is only linear in limited regions. Its gain is given by Equation 2.4. Depending on the input phase, K_{PD} can be either positive or negative.

$$K_{PD} = \pm \frac{V_{dd}}{\pi} \tag{2.4}$$

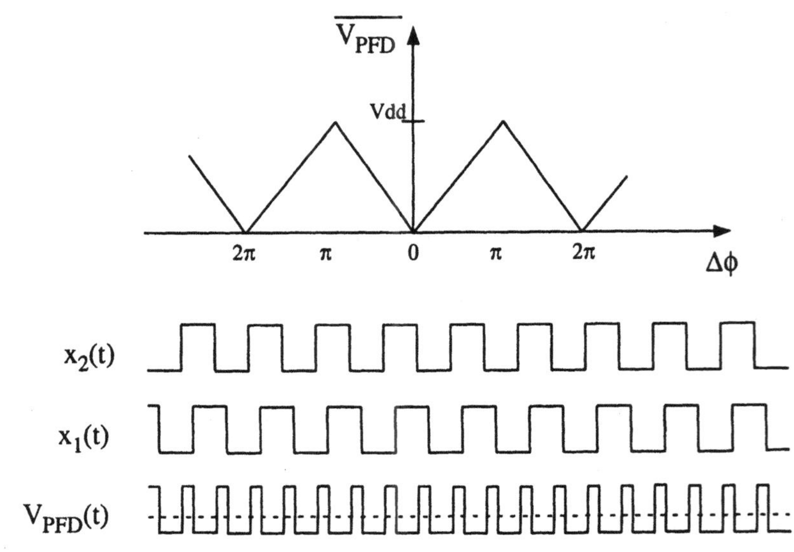

Figure 2.4. Characteristics of an XOR gate as a PD

More often a phase/frequency detector (PFD) as shown in Figure 2.5 is used in combination with a charge pump. It detects the phase difference for two inputs at the same frequency. It also detects the frequency difference if the inputs have the different frequencies. This significantly increases the acquisition range and speeds up the phase lock process of PLLs. In contrast to multipliers and XOR PD where an output voltage is generated directly, a PFD in Figure 2.5 generates non-complementary outputs *UP* and*DOWN* which control the charge pump.

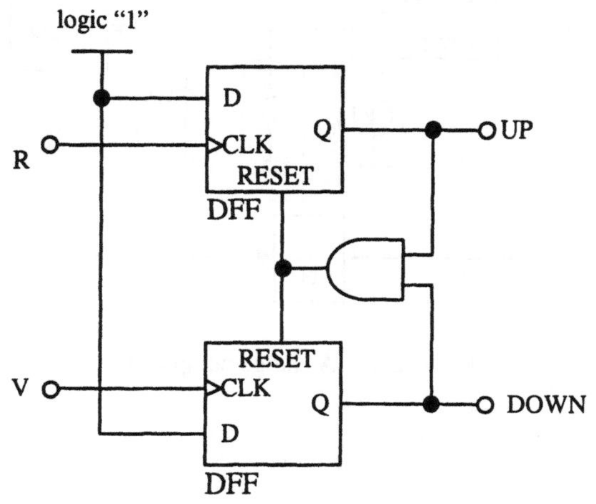

Figure 2.5. A phase/frequency detector (PFD)

A PLL with a PFD and a charge pump is usually called a charge pump PLL. A typical charge pump PLL diagram is drawn in Figure 2.6. The outputs of the PFD are used to turn on and off the switches S_1 and S_2, which control the up and down currents I_{CP}. Example outputs of the PFD are shown in Figure 2.7. When $f_R < f_V$ as shown in (a), the *DOWN* signal is turned high whenever a rising edge of *V* is generated and is turned low whenever a rising edge of *R* is generated. Meanwhile, the *UP* signal remains low. This discharges the capacitors in the loop filter and slows down the oscillation of the VCO. Hence it realizes frequency detection. When *R* and *V* are at the same frequency, but *R* is taking a phase lead as shown in (b), the *UP* signal is turned high periodically while the *DOWN* signal remains low. This charges up the capacitors in the loop filter and speeds up the oscillation of the VCO.

At the beginning of the phase lock acquisition process, the *R* and *V* signals are likely to be at different frequencies. With the help of the PFD, the negative feedback loop adjusts the VCO frequency and brings the frequency of *V* close to the frequency of *R*. Then the PFD functions as a PD. The phase error information generated by the PFD and the loop filter is used to smooth out the high-frequency component from the charge pump. The DC average value is used to fine tune the phase for the VCO. The phase lock condition is not satisfied until both the frequency and the phase of the output clock are synchronized with the input clock.

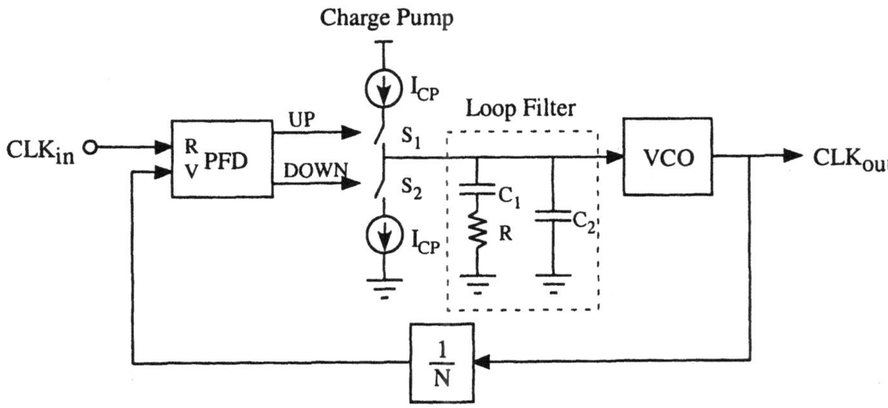

Figure 2.6. A charge pump PLL

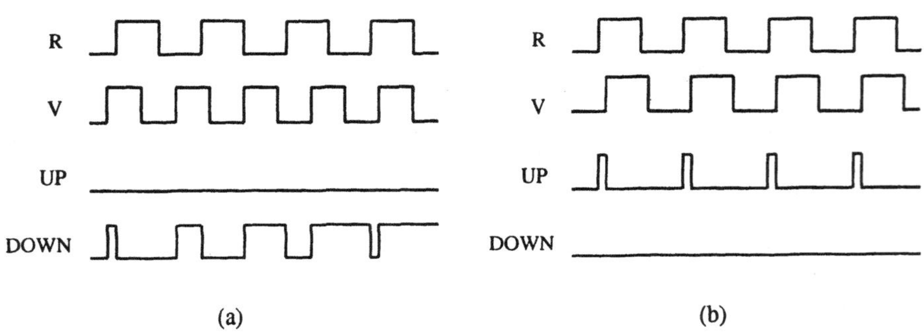

Figure 2.7. Example inputs and outputs for PFD (a) $f_R < f_V$ (b) V is lagging R

The phase lock acquisition is a highly nonlinear process because of the non-linear function of the PFD. It switches its function from frequency detection to phase detection. It is only monotonic within a certain phase range, beyond which the phase rolls over and its characteristic becomes non-monotonic.

There is another process in a PLL called *tracking*. Once the phase lock is acquired, the phase error is usually retained in a limited range. Hence the PD or the PFD behaves as a linear device and the PLL becomes a fairly linear system.

In the phase-locked state, the PFD and the charge pump are used as a PD whose inputs are the two phases and the output is a current. We take an example where V is lagging R as the case in Figure 2.7 (b). If the period for R and V is T and the phase difference between R and V is $\Delta\phi$, then the on-time for switch

S_1 is

$$t_{UP} = \frac{\Delta\phi T}{2\pi} \tag{2.5}$$

during one period, and switch S_2 is always off. When both switches are off, the charge pump output is held at high impedance and there is no current output. So the average output current is given by

$$\overline{I_{PD}} = I_{CP}\frac{t_{UP}}{T} = \frac{I_{CP}}{2\pi}\Delta\phi \tag{2.6}$$

It can be shown that Equation 2.6 is also valid for the case where R is lagging V. In that case, both the phase error $\Delta\phi$ and the average output current $\overline{I_{PD}}$ are negative. Therefore, when the PFD and the charge pump are used together as a PD, its gain is given by

$$K_{PD} = \frac{I_{CP}}{2\pi} \tag{2.7}$$

However, there can be a potential problem for the PFD in Figure 2.5. As the phase error approaches zero, which is the ideal case for the locked state, the pulses at *UP* or *DOWN* become infinitely narrow such that the logic gates cannot make a complete transition due to the finite rise and fall times. Therefore, for small phase errors, the switches S_1 and S_2 cannot be turned on completely effectively, and the PFD combined with the charge pump can no longer generate an average current that is proportional to the input phase difference. This phenomenon is shown in Figure 2.8. This results in a PFD gain which is close to zero for small phase errors. This is illustrated in Figure 2.9. The input phase range with a reduced gain is called a *dead zone*.

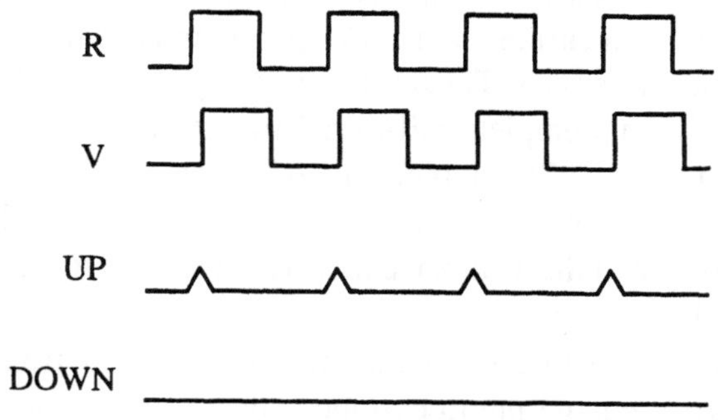

Figure 2.8. Illustration of dead zone

The dead zone of a PFD is extremely harmful for a PLL in terms of phase noise, since it occurs right in the middle of the locked condition. When the

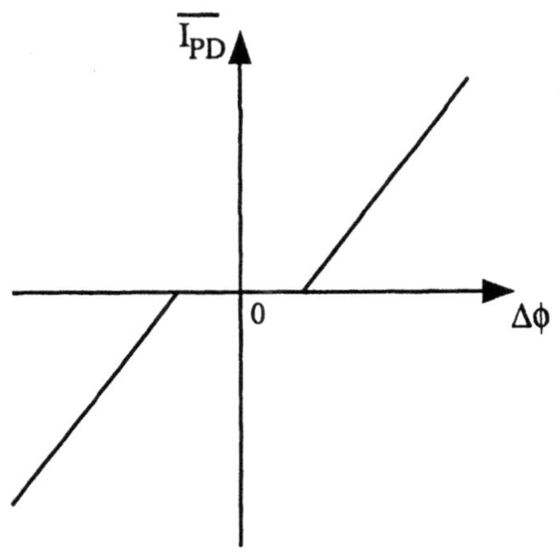

Figure 2.9. PD characteristics with dead zone

phase error falls in the dead zone, the PFD fails to feedback the phase error information which is used for correcting the VCO phase error. As a result, the output phase can drift anywhere within the dead zone without control. The VCO is essentially operated open-loop in the dead zone.

There has been a lot of research effort to eliminate the dead zone for a PD. A well-known solution is to add some additional delay in the NAND gate in Figure 2.5. Instead of having the PFD in a high impedance state as the phase error approaches zero, such a PFD generates pulses at *UP* and *DOWN* with equal width. As a result, both switches $S1$ and $S2$ are kept on for an equal period of time during each clock cycle, and they remain off for the rest of the period. When the up and down currents in the charge pump have the same magnitude, the net output current is zero. This is shown in Figure 2.10 (a). The pulses are wide enough such that complete rail-to-rail transitions are made. When there is finite phase error at the PFD inputs, one of the output pulses becomes wider than the other. As a result, there is a net up or down current. This is illustrated in Figure 2.10 (b). With the dead zone removed, the PFD maintains its nominal gain even in the locked state.

It is also worthwhile to discuss the effects that result from the mismatch of the up and down charge pump currents, since it is an important issue in a charge pump PLL design. Without any current mismatch, in the locked state, the PFD generates narrow pulses with equal width, and the net output current is zero. However, if any current mismatch exists, one of the switches of $S1$ and $S2$ in Figure 2.6 has to be turned on for a longer period of time than the other. Without losing generality let us assume the the current flowing down from V_{dd} is slightly larger than the current flowing into ground. When pulses

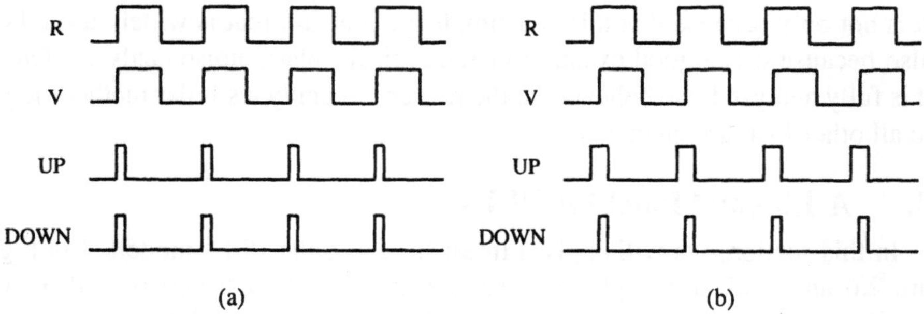

Figure 2.10. Elimination of dead zone (a) with zero phase error (b) with finite phase error

with equal widths are generated on *UP* and *DOWN* signals, there is a net up current on average during each clock cycle. This results in an increased control voltage for the VCO which pushes the VCO phase forward. Then the PFD will generate slightly wider DOWN pulses such that the average output current from the charge pump in a cycle is zero. Now the PLL is in the locked state and the situation is illustrated in Figure 2.11. However, the instantaneous output current is not always zero in this case. It generates periodic current pulses which cause ripple at the VCO control at the same frequency as the PFD update rate. The VCO is frequency modulated by the ripple. As a result, spurious tones at an offset equal to the input reference frequency and its harmonics can be seen at the PLL output spectrum. This phenomenon is usually called *reference spur generation.* It is a serious problem in many applications and has to be minimized in a design.

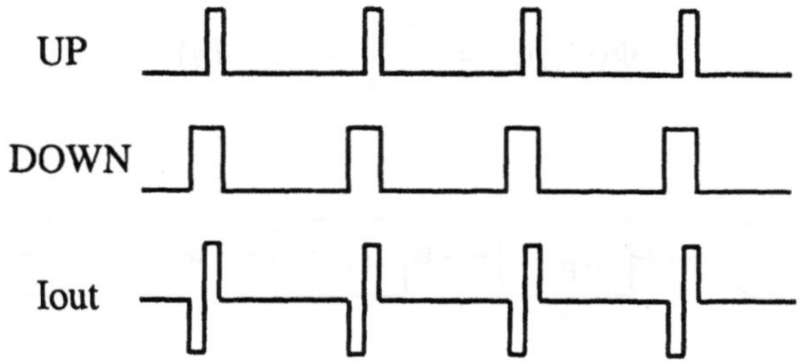

Figure 2.11. Periodic output current from the charge pump due to current mismatch

In this book, we focus on the phase noise analysis which is only valid after the phase lock is established. Hence we apply a linear model to the PLL. In particular, we are interested in the charge pump PLL as drawn in Figure 2.6.

It is not only because that this is a simple architecture that is widely used, but also because it is a good example PLL circuit for phase noise analysis. Once it is fully analyzed, we believe that the reader can apply a similar methodology to all other PLL architectures.

3. A Linear Model for PLLs

In this section, we will apply a linear model to the PLLs modelled in Figure 2.6 and analyze the phase noise contribution of each part of a PLL. In particular, we will focus on charge pump PLLs which are widely used in many applications.

A linear model for the PLLs is drawn in Figure 2.12. The PD is characterized as

$$V_{PFD} = K_{PD}\Delta\phi \tag{2.8}$$

where K_{PD} is a constant called the PD gain, and $\Delta\phi$ is the phase difference between its two inputs. $G(s)$ is the transfer function of the loop filter. The VCO is modelled as an integrator. The reason is as follows. The transfer characteristic of a VCO can be written as

$$f_{VCO}(t) = K_{VCO}V_{cntrl}(t) \tag{2.9}$$

where V_{cntrl} is the input control voltage and K_{VCO} is a constant factor called the VCO gain. Integrating Equation 2.9 on both sides, we get

$$\phi_{VCO}(t) = K_{VCO}\int_0^t V_{cntrl}(t)\,dt \tag{2.10}$$

This corresponds to the s-domain expression as in Equation 2.11.

$$\Phi_{VCO}(s) = \frac{K_{VCO}}{s}V_{cntrl}(s) \tag{2.11}$$

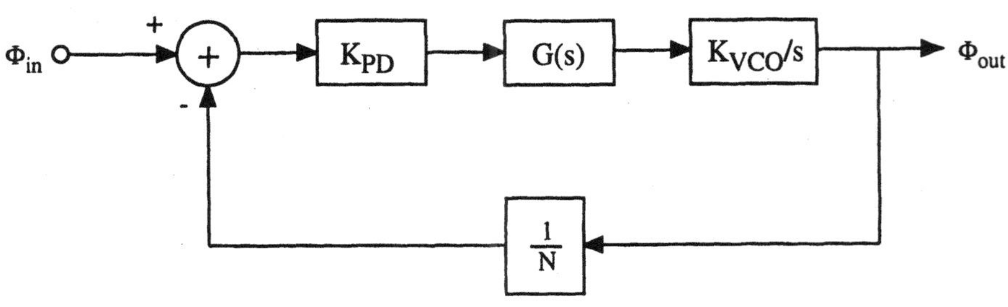

Figure 2.12. A linear model for a PLL in locked state

Based on the block diagram in Figure 2.12, the closed-loop transfer function of the PLL can be derived as in Equation 2.12.

$$\begin{aligned} H(s) &= \frac{\Phi_{out}(s)}{\Phi_{in}(s)} \\ &= \frac{K_{PD}K_{VCO}G(s)}{s+\frac{1}{N}K_{PD}K_{VCO}G(s)} \end{aligned} \tag{2.12}$$

Charge pump PLLs, as shown in Figure 2.6, are commonly used in many applications, because they have a number of advantages over other types of PLLs. The PFD in the charge pump PLLs helps increase the lock range and speed up the capture process. The charge pump combined with the loop filter forms an integrator with infinite gain at DC. This results in zero phase error for an ideal locked case. Therefore, let us take the charge pump PLLs as an example to analyze the noise contribution from each part in the PLL.

We have derived the linearized gain for the PD in the previous section. The loop filter transfers the charge pump output current into a low-pass filtered voltage.

For a charge pump PLL, the parameters in Figure 2.12 are given by Equations 2.13 and 2.14, where I_{CP} is the charge pump current and R, C_1 and C_2 are loop filter components as shown in Figure 2.6.

$$K_{PD} = \frac{I_{CP}}{2\pi} \tag{2.13}$$

$$G(s) = \frac{1 + sRC_1}{s\,(C_1 + C_2 + sRC_1C_2)} \tag{2.14}$$

In order to simplify our analysis, we first neglect C_2, since it is usually much smaller than C_1. It is easier to analyze the loop without C_2 and then explain why C_2 has to be added. Now the loop filter transfer function is simplified to

$$G(s) = \frac{1 + sRC_1}{sC_1} \tag{2.15}$$

We are interested in the noise contribution from the input clock, the charge pump, the loop filter, the VCO and the feedback frequency divider. A linear model for the charge pump PLL is drawn in Figure 2.13 with the noise sources added.

The open-loop and closed-loop transfer functions for Figure 2.13 can be found as

$$H_{open}(s) = \frac{K'_{PD}K_{VCO}(sRC_1 + 1)}{s^2C_1} \tag{2.16}$$

and

$$H(s) = \frac{K'_{PD}K_{VCO}R(s + \frac{1}{RC_1})}{s^2 + sK'_{PD}K_{VCO}R + \frac{K'_{PD}K_{VCO}}{C_1}} \tag{2.17}$$

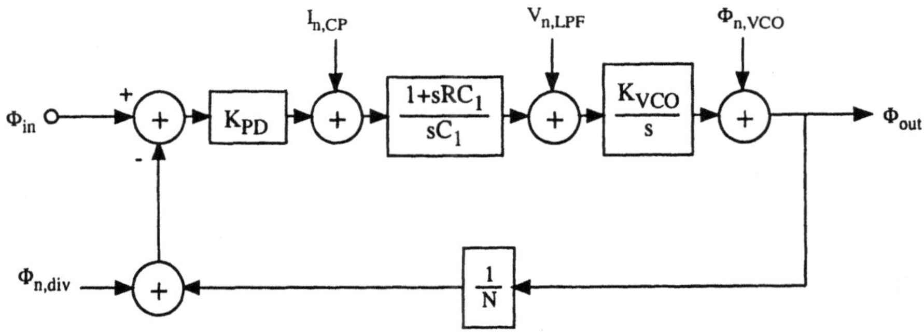

Figure 2.13. A linear model for a charge pump PLL with noise sources

where $K'_{PD} = \frac{K_{PD}}{N} = \frac{I_{CP}}{2\pi N}$ [Wilson et al., 2000]. Equation 2.12 indicates that the PLL has a DC gain of N. This is consistent with the phase-lock condition.

The charge pump charging and discharging the capacitor C_1 implements an integrator. The VCO is modelled as another integrator as we discussed before. Hence the open-loop transfer function has a double pole at DC. This makes the loop inherently unstable, since with 180° phase shift given by the two poles at DC and a gain larger than unity, the loop can easily oscillate. So the resistor R has to be added to introduce a zero in order to stabilize the loop.

Comparing the denominator for Equation 2.12 and the generalized form for a second order system

$$s^2 + 2\zeta\omega_n s + \omega_n^2,$$

the natural frequency ω_n and the damping factor ζ can be derived as

$$\omega_n = \sqrt{\frac{K'_{PD} K_{VCO}}{C_1}} \tag{2.18}$$

and

$$\zeta = \frac{R}{2}\sqrt{K'_{PD} K_{VCO} C_1} \tag{2.19}$$

It is seen from Equation 2.19 that the purpose of the resistor R is to increase the damping factor ζ so that the loop is stabilized. Without the resistor, ζ becomes 0, and the closed-loop function has two poles on the imaginary axis. Therefore, the loop is unstable.

In the phase locked state when the switches S_1 and S_2 are turned on during the short pulses, there could be a net output current from the charge pump due to the mismatch between the up and down currents. The charge pump output current generates voltage ripples across the resistor R. This results in frequency modulation (FM) of the VCO output. Spurs at the offset frequencies equal to the input reference frequency and its harmonics can be observed at the PLL output spectrum.

Thus a small capacitor C_2 is usually added to reduce the voltage ripple at the VCO control terminal. It adds another pole to both the PLL open-loop and the closed-loop transfer functions, and further filters out the high-frequency spurs and noise. Adding at zero at ω_z and a pole at ω_p with $\omega_z < \omega_p$ for loop stabilization is usually called *lead-lag compensation*.

The open-loop transfer function is illustrated in Figure 2.14. The dashed line is the transfer function without the resistor R and the capacitor C_2. Due to the double pole at DC, it has -180° phase shift at the $0dB$ crossover frequency K_0 which causes the loop to be unstable. The zero at ω_z and the pole at ω_p are given by

$$\omega_z = \frac{1}{RC_1} \tag{2.20}$$

$$\omega_p = \frac{C_1 + C_2}{RC_1C_2} \approx \frac{1}{RC_2} \tag{2.21}$$

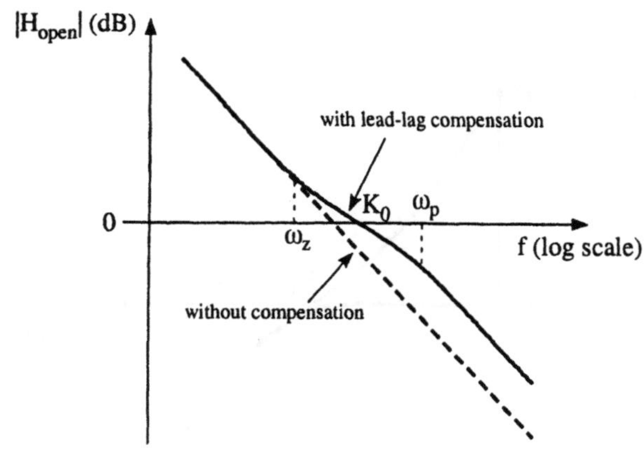

Figure 2.14. A PLL open-loop transfer characteristic with and without lead-lag compensation

With the help of the zero at ω_z, the open-loop phase response is close to -90° instead of -180° at frequency K_0. The loop is stabilized with an increased phase margin. Please notice that both ω_z and ω_p have to be located far away enough from K_0 to maintain good phase margin, since both of them are trying to pull the phase response back to -180° at frequency K_0. If ω_z and ω_p are too close to K_0, it can result in overshoot in the closed-loop response.

From Equation 2.12 it is seen that the transfer function from the PLL input to the output has a low-pass characteristic. The $-3dB$ bandwidth for the closed-loop transfer function is the same as the $0dB$ bandwidth for the open-loop transfer function K_0. If we assume that both ω_z and ω_p are far enough from K_0, then the loop filter looks like a single resistor R at frequency K_0. The open-loop transfer function can then be approximated by Equation 2.22. Hence the loop bandwidth is given by Equation 2.23 which is sometimes called open-loop gain.

$$H_{open}(s)|_{s \approx jK_0} \approx \frac{K_{PD}K_{VCO}R}{Ns} = \frac{K'_{PD}K_{VCO}R}{s} \tag{2.22}$$

$$K_0 = K'_{PD}K_{VCO}R = 2\zeta\omega_n \tag{2.23}$$

Figure 2.15 shows how to obtain the closed-loop transfer function from the open-loop transfer function. The closed-loop curve stays close to $0dB$ for frequencies lower than K_0, and follows the closed-loop curve for frequencies higher than K_0. As a result, the open-loop pole ω_p is also a closed-loop pole, and it helps reduce the high-frequency phase noise in a PLL. Therefore, the frequency for ω_p should be lowered as much as possible without disturbing the loop stability. An empirical optimal value for ω_p is given by $\omega_p = 4K_0$ [Wolaver, 1991].

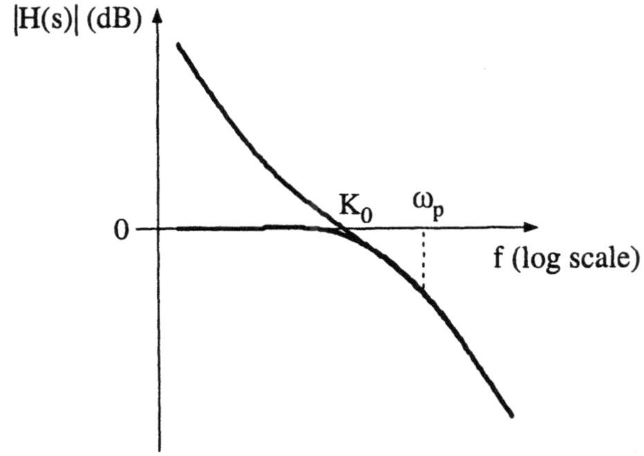

Figure 2.15. Comparison of closed-loop and open-loop response for a charge pump PLL

Similarly we can derive the noise transfer function from each block in the PLL to the output while neglecting C_2. In particular, we are interested in the noise generated in the PFD, the loop filter, the VCO and the frequency divider. The transfer functions are derived as follows where ω_n and ζ are given by Equations 2.18 and 2.19.

$$H_{PD}(s) = \frac{\Phi_{out}(s)}{I_{n,CP}(s)} = \frac{K_{VCO}}{C_1} \frac{sRC_1 + 1}{s^2 + 2\zeta\omega_n s + \omega_n^2} \quad (2.24)$$

$$H_{LPF}(s) = \frac{\Phi_{out}(s)}{V_{n,LPF}(s)} = \frac{K_{VCO}s}{s^2 + 2\zeta\omega_n s + \omega_n^2} \quad (2.25)$$

$$H_{VCO}(s) = \frac{\Phi_{out}(s)}{\Phi_{n,VCO}(s)} = \frac{s^2}{s^2 + 2\zeta\omega_n s + \omega_n^2} \quad (2.26)$$

$$H_{div}(s) = \frac{\Phi_{out}(s)}{\Phi_{n,div}(s)} = \frac{K_{PD}K_{VCO}}{C_1} \frac{sRC_1 + 1}{s^2 + 2\zeta\omega_n s + \omega_n^2} \quad (2.27)$$

It is seen from Equation 2.24 that the transfer function from the PFD to the PLL output is the same as the transfer function from the input clock except for a gain factor for the PFD. This means that it is also low-pass filtered with a bandwidth of K_0 and it has another pole at frequency ω_p.

Equation 2.25 indicates that the noise from the loop filter is bandpass filtered, since its only zero is moved to DC. Its passband is centered around ω_n given by Equation 2.18.

The phase noise due to the VCO is the major noise contributor in many applications, and its transfer function is given by Equation 2.26. The way to derive its frequency response curve is demonstrated in Figure 2.16.

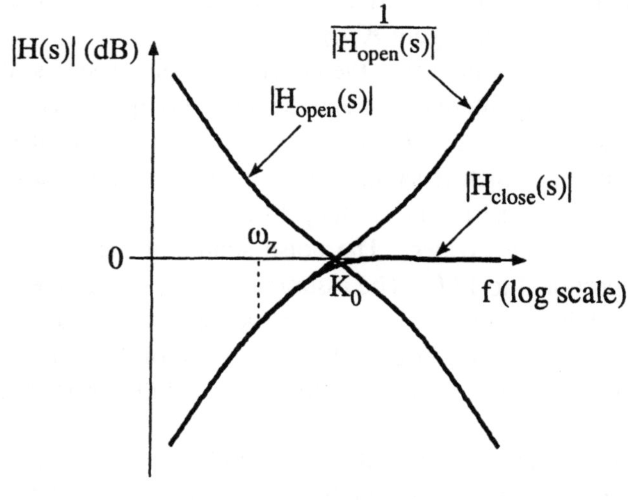

Figure 2.16. Derivation of noise transfer characteristic for VCO

The relation between the closed-loop transfer function from the VCO to the PLL output and the open-loop transfer function is given by

$$\begin{aligned} H_{close}(s) &= \frac{1}{1+H_{open}(s)} \\ &\approx \begin{cases} \frac{1}{H_{open}(s)} & for|H_{open}(s)| \gg 1 \\ 1 & for|H_{open}(s)| \ll 1 \end{cases} \end{aligned} \quad (2.28)$$

In Figure 2.16, the curve for $H_{open}(s)$ is first flipped vertically with respect to the x-axis to get the curve for $1/H_{open}(s)$. The curve for $H_{close}(s)$ follows the curve $1/H_{open}(s)$ for frequencies lower than K_0 and is approximately equal to one for frequencies higher than K_0. It is seen from the plot that the phase noise from the VCO propagates to the PLL output without attenuation for $\omega > K_0$. Now the zero ω_z in the open-loop transfer function becomes a pole in the closed-loop transfer function. Hence the VCO noise sees a first order attenuation for $\omega_z < \omega < K_0$, and sees a second order attenuation for $\omega < \omega_z$. Therefore, the open-loop zero helps reduce the VCO noise at low frequencies. It is beneficiary for the sake of phase noise to move ω_z to a higher frequency. However, when it gets too close to K_0, the phase margin is reduced, and this jeopardizes the loop stability. It can cause unwanted peaking in the frequency response or even cause the loop to be unstable. An empirical optimal distance between ω_z and K_0 is $\omega_z = (1/4)K_0$ [Wolaver, 1991].

It is seen from Equation 2.26 and Figure 2.16 that the VCO phase noise is high-pass filtered before it reaches the PLL output, and the corner frequency is K_0 — normally called the PLL *loop gain* or *loop bandwidth* .

Equation 2.27 suggests that the noise generated from the frequency divider is also low-pass filtered. The difference between Equation 2.27 and 2.24 is only a constant factor of the PFD gain K_{PD}.

It is reasonable to assume that the equivalent noise sources at the charge pump current, the loop filter voltage and the divider output phase are white and Gaussian. Practical causes of these noise sources might include the thermal noise in the current which is sampled to the output of the charge pump, the thermal noise of the resistor in the loop filter, as well as the device thermal noise in the frequency divider. The noise source at the VCO has a power spectrum proportional to $1/f^2$. The assumptions employed here ignore flicker noise for simplicity. Flicker noise has a power spectrum of $1/f$, and is often a significant source of noise in CMOS PLLs. Each of these noise sources are then scaled by their respective transfer functions and their power is summed at the PLL output.

When the noise power spectral density is multiplied by the power gain and integrated over frequency, we obtain the power of the output jitter. We hold the following assumptions:

$$I_{n,CP}(s) = \overline{i_n^2}/\Delta f \quad (2.29)$$

$$V_{n,LPF}(s) = \overline{v_n^2}/\Delta f \tag{2.30}$$
$$\Phi_{n,VCO}(s) = K_\theta / f^2 \tag{2.31}$$
$$\Phi_{n,div}(s) = \overline{\phi_n^2}/\Delta f \tag{2.32}$$

Now we integrate the output noise power over the frequency and obtain the following results for the output phase noise due to $I_{n,CP}$, $V_{n,LPF}$, $\Phi_{n,VCO}$ and $\Phi_{n,div}$:

$$\phi_{n,CP}^2 = \frac{1}{2\pi}\int_0^\infty (\overline{i_n^2}/\Delta f)\,|H_{PD}(j\omega)|^2\, d\omega = \frac{(\overline{i_n^2}/\Delta f)\omega_n}{2\pi K_{PD}'^2}\Gamma_{CP}(\zeta) \tag{2.33}$$

$$\phi_{n,LPF}^2 = \frac{1}{2\pi}\int_0^\infty (\overline{v_n^2}/\Delta f)\,|H_{LPF}(j\omega)|^2\, d\omega = \frac{(\overline{v_n^2}/\Delta f)K_{VCO}^2}{2\pi\omega_n}\Gamma_{LPF}(\zeta) \tag{2.34}$$

$$\phi_{n,VCO}^2 = \frac{1}{2\pi}\int_0^\infty \frac{K_\theta}{f^2}\,|H_{VCO}(j\omega)|^2\, d\omega = \frac{2\pi K_\theta}{\omega_n}\Gamma_{VCO}(\zeta) \tag{2.35}$$

$$\phi_{n,div}^2 = \frac{1}{2\pi}\int_0^\infty (\overline{\phi_n^2}/\Delta f)\,|H_{div}(j\omega)|^2\, d\omega = \frac{(\overline{\phi_n^2}/\Delta f)N^2\omega_n}{2\pi}\Gamma_{CP}(\zeta) \tag{2.36}$$

where Γ_{CP}, Γ_{LPF} and Γ_{VCO} are only dimensionless functions of ζ.

One of the common techniques to reduce the PLL output phase noise is to increase the loop natural frequency ω_n, or equivalently the loop bandwidth K_0, by increasing K_{PD}. It is seen from Equation 2.23 that ω_n can be increased by the same factor as K_{PD} while K_{VCO} is kept constant. Equations 2.33 through 2.36 suggest that this would reduce the noise from the charge pump, loop filter and the VCO. The noise from the frequency divider will be increased, fortunately, the divider contributes negligible noise in most cases. It only becomes significant in the extreme high-frequency applications in the GHz range where the clock transition times take a considerable portion of the clock period.

There are many applications where the input is a clean clock source with negligible phase noise such as in the case of a frequency synthesizer. Then the loop bandwidth should be maximized to minimize the phase noise. There are also cases where the there is significant amount of jitter in the input clock that has to be filtered out by the PLL. When the loop bandwidth is increased to minimize the phase noise generated by the PLL, much of the phase noise for the input clock will transfer to the output. Therefore, sometimes an optimal loop bandwidth has to be chosen to minimize the total output phase noise.

On the other hand, increasing K_0 infinitely may cause the loop to be unstable. The analysis that we have performed so far is based on a continuous-time approximation. The charge pump is considered as a linear device with a gain of $I_{CP}/(2\pi)$. However, a charge pump is really a discrete-time component. The phase comparison is done only at the clock transitions, and the output is generated in the form of current pulses. The maximum loop bandwidth is restricted by the update rate of the PFD. In general, it has approximately to be less than 1/10 of the PFD update rate to avoid instability. The loop bandwidth has to be further limited when there is significant phase noise in the input clock that needs to be attenuated.

Once a desirable K_0 has been chosen, the phase noise can be further reduced by increasing K_{PD} and reducing K_{VCO} by the same factor. This also reduces the noise from the charge pump and the loop filter.

It is also seen from Equations 2.33 and 2.36 that the noise power from the charge pump and the feedback divider is amplified by a factor of N^2. As we discussed before, the phase noise in the input clock is also amplified by N^2. This can potentially become a problem for PLLs with large divider ratios. In particular, the charge pump usually dominates the close-in phase noise if N is large.

When a ring oscillator is used in a CMOS design, its phase noise is the dominant contributor outside the PLL loop bandwidth. Since there is no frequency selective element in a ring oscillator, its phase noise can be very large. In this book, we will focus on phase noise theories and practical designs of low-phase-noise CMOS ring oscillators.

4. Conclusions

In this chapter we have reviewed some basic knowledge for PLLs. In particular, we have studied the noise transfer functions from each of the building blocks in a charge pump PLL. We have proved that the noise from the input clock, the charge pump and the feedback frequency divider is low-pass filtered before it reaches the PLL output. The noise from the resistor in the loop filter is bandpass filtered. The phase noise generated by the VCO is high-pass filtered.

In order to minimize the phase noise, the loop bandwidth should be maximized by maximizing the PD gain. This should be done without jeopardizing the loop stability.

Chapter 3

PHASE NOISE AND TIMING JITTER

Phase noise and timing jitter are closely related terms. Their values can sometimes be derived from each other. Before further analyzing the oscillator phase noise, it is necessary to look closely at the definitions for both terms and understand their relationship.

1. Phase Noise

An ideal sinusoidal waveform can be mathematically written as

$$V(t) = A\sin(\omega t) \tag{3.1}$$

Both its amplitude and phase can be corrupted by noise. Hence a noisy waveform can be written as

$$V(t) = A(t)\sin[\omega t + \phi(t)] \tag{3.2}$$

The amplitude noise and the phase noise are treated differently in practical systems. The amplitude noise can be removed easily by a hard limiter, while the phase noise, once introduced, is likely to remain. There is usually some amplitude noise converted into phase noise when the signal is processed by a limiter. However, this will be neglected in our analysis, since its effect is negligible in many cases, and taking it into consideration makes the analysis too cumbersome.

Each periodic waveform can be expanded into sinusoidal waveforms by Fourier series. Our argument below can be interpreted as its first harmonic. Our analysis can also be easily extended to cover waveforms with any periodic shapes.

The impact of phase noise is illustrated in Figure 3.1 in both time domain and frequency domain. The spectrum of a sinusoidal waveform is a δ-function.

Once its phase is corrupted by noise, the waveform loses its periodicity. Hence its spectrum is no longer a δ-function. Instead it has a finite span in the frequency domain as shown in the figure.

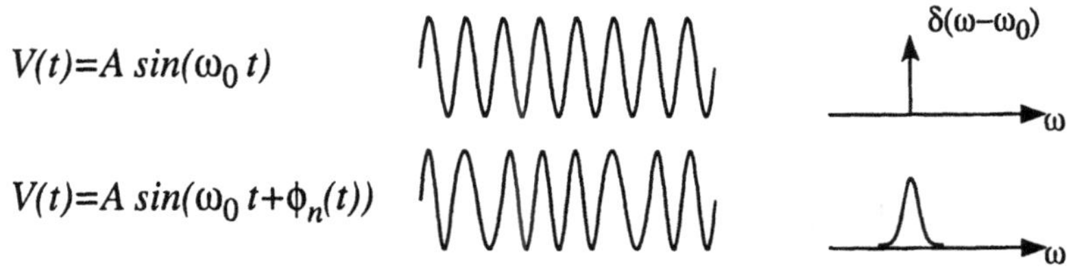

Figure 3.1. Illustration of phase noise and timing jitter

Wireless communications systems usually impose stringent requirements on the phase noise of frequency synthesizers. Each communications channel only occupies a narrow bandwidth and they are allocated next to each other with small spacings. The excessive phase noise generated by a frequency synthesizer can spread its energy over to the adjacent channels. There are situations where the signal in the adjacent channel is much stronger than the signal in the desired channel. When mixed with the phase noise spread over to the adjacent channel, the interferer can be converted into the same channel as the desired signal. This results in cross talk.

A commonly used term is single-sideband (SSB) phase noise. Its definition can be referred to Figure 3.2 where the power spectral density (PSD), $S(\omega)$, of a sine wave is shown. The SSB phase noise is defined as

$$L(\Delta\omega) = \frac{S(\omega_0 + \Delta\omega)}{\int_{-\infty}^{+\infty} S(\omega)\, d\omega} \qquad (\Delta\omega > 0) \tag{3.3}$$

Because the PSD is symmetrical with respect to the center frequency ω_0, Equation 3.3 is only defined for $\Delta\omega > 0$. It is obvious to see that the SSB phase noise, $L(\Delta\omega)$, satisfies

$$\int_0^{+\infty} L(\Delta\omega)\, d\Delta\omega = \frac{1}{2} \tag{3.4}$$

Figure 3.2 does not represent the actual shape of a SSB phase noise spectrum for a CMOS VCO. It is only for illustration purposes. The actual shape of SSB phase noise often resembles the one shown in Figure 3.3. The spectrum can be divided into three regions: a $1/f^3$ region, a $1/f^2$ region and a flat region [Leeson, 1966, Hajimiri and Lee, 1998]. If white noise is the only noise source that exists in a oscillator, it can be derived from different approaches

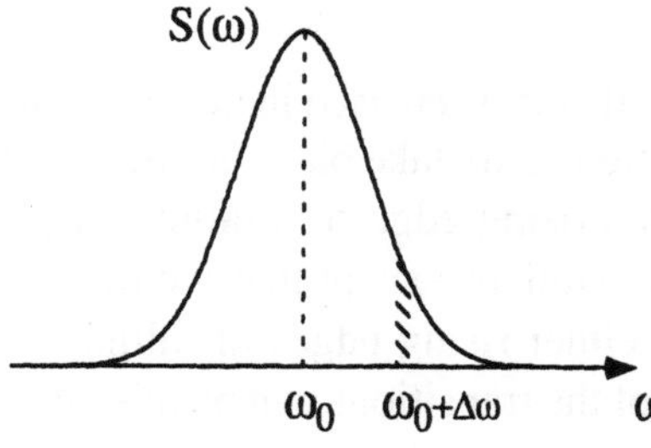

Figure 3.2. Definition of SSB phase noise

that the phase noise is proportional to $1/\Delta f^2$. This will be discussed in greater details when we provide a brief review of the existing phase noise theories in the next chapter. MOS devices have flicker noise whose power spectral density is proportional to $1/f$. It causes the phase noise to rise at a faster rate at low frequencies. This results in the $1/f^3$ region. Both the $1/f^3$ region and the $1/f^2$ region are due to the device noise within the resonator or the ring of the ring oscillator. When the signal is used to drive the next stage, an output buffer is usually needed. It not only amplifies the internal voltage and sends it to the output, but also helps prevent load pulling or the variation of frequency due to any change in the output load. If there is any white noise in the output buffer or the measurement equipment, it causes the flat region in the phase noise spectrum.

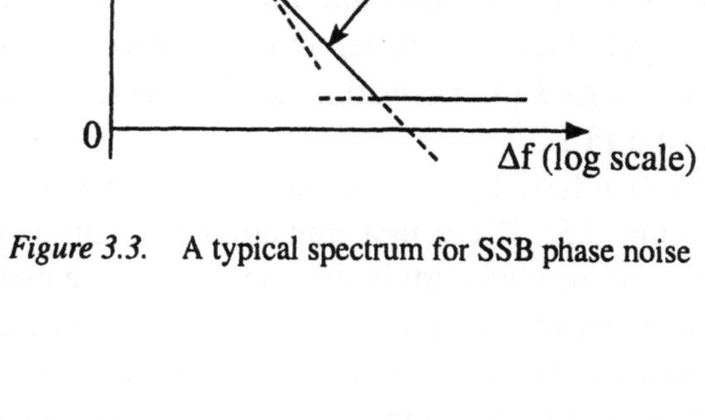

Figure 3.3. A typical spectrum for SSB phase noise

2. Timing Jitter

Timing jitter is a term that is often used in data communications. For a digital clock, ideally, the transitions only take place at time $t = kT_c$ or $t = (k+1/2)T_c$ depending whether it is a rising edge or a falling edge, where T_c is the clock period. Data streams are similar, except that the transitions only occur at time $t = kT_c$. They can be either rising edges or falling edges, depending on the current state, and some of the transitions can be missing if two or more identical symbols arrive one after another.

With the presence of circuit noise, the transitions of the clock or data takes place either earlier or later than the ideal transition times. This is called timing jitter. It can cause the data to be misaligned with the clock in time so that the data symbols are not sampled at the optimal points. It can be possibly sampled during its transition or even at the adjacent symbol. This results in increased bit error rate (BER) and can be a serious problem in high-speed data communications.

Jitter describes the variation in time of the clock or data transitions. Even though the definition for jitter sounds straightforward, it can be confusing or misleading without mentioning the timing reference. The question is: when we talk about timing variation, what is the *ideal* clock as a timing reference? This is directly related to the way that jitter is measured.

The way of measuring the timing jitter of a PLL can be summarized into two categories: *self-referred jitter* and *input-referred jitter*. Input referred jitter is also sometimes called *absolute jitter* [McNeill, 1997]. Both forms are shown in Figure 3.4. In Figure 3.4 (a), the device under test (DUT) is a stand-alone VCO or a PLL that is locked to another clock source. The signal analyzer is triggered by the PLL/VCO output clock by itself. One clock transition that triggers the signal analyzer is taken as a reference point, and the jitter of another transition in the same clock is measured. In Figure 3.4 (b), the PLL is phase-locked onto the input, and its output clock is sent to a signal analyzer such as an oscilloscope which measures the jitter. The source clock is used to trigger the signal analyzer. In other words, the output phase of the PLL is compared with the input reference clock and that is why it is called input-referred jitter.

Interestingly the self-referred jitter and input-referred jitter for a PLL that is locked onto the input clock have different characteristics. For a stand-alone oscillator, only self-referred jitter is valid, and it behaves differently from the self-referred jitter of a PLL.

Let us consider self-referred jitter of a stand-alone oscillator first. This case is illustrated in Figure 3.5. If the first transition is taken as a reference, the following transitions occur with certain uncertainty in time due to the presence of circuit noise as shown in the figure. If the noise is white and Gaussian, it is a random walk process from one transition to the next. If we only consider the rising edges of the clock (the falling edges can be considered similarly), the

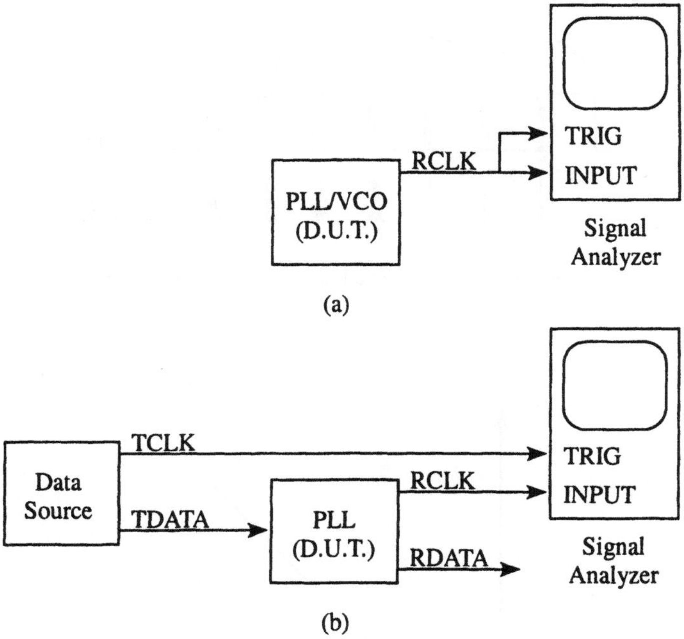

Figure 3.4. Jitter measurement (a) self-referred (b) input-referred

nominal time interval is T_c. The timing jitter causes the time interval to have a variance of σ^2. Since the random process of going from one transition to the next is uncorrelated from other time intervals, the jitter of the nth rising edge after the reference edge is characterized by a variance of $n\sigma^2$. This proportionality of jitter power with respect to time can be generalized by Equation 3.5 where τ is the time interval between the reference edge and the edge whose jitter is being measured. Now the self-referred jitter can be characterized by a single parameter c with a unit of time. If the self-referred jitter is measured on different edges, a plot as shown in Figure 3.6 can be obtained. The parameter c can be obtained from the slop of the straight line. In particular, the jitter of one period $\sigma^2(T_c)$ is sometimes called *period jitter.*

$$\sigma^2(\tau) = c\tau \tag{3.5}$$

Next let us look at the self-referred jitter for a PLL. The phase noise due to the VCO is suppressed within the loop bandwidth K_0. As a result, when the time interval $\tau < 1/K_0$, the negative feedback loop does not take effect and its jitter is the same as that of a stand-alone VCO. As the time interval increases such that $\tau > 1/K_0$, the timing correction by the negative feedback takes effect.

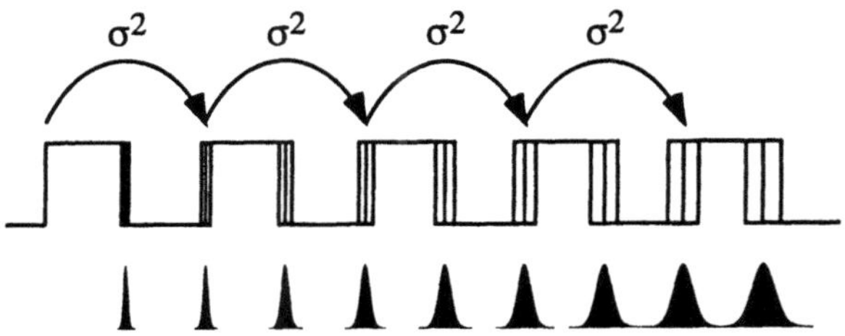

Figure 3.5. Self-referred jitter of a stand-alone VCO

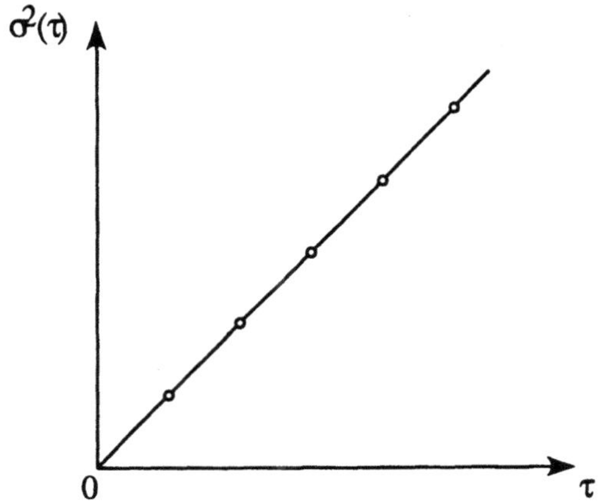

Figure 3.6. Self-referred jitter vs. time interval for a VCO

Therefore, the jitter power stops increasing and saturates to a limited value. This is shown in Figure 3.7. The self-referred jitter flattens out at the level of $\sigma_s^2 = c/K_0$.

Finally we will study the input-referred jitter of a closed-loop PLL. We assume that the input clock to be a clean source without jitter. Let us refer to a charge pump PLL in Figure 2.6 in the last chapter, assuming $N = 1$ for simplicity. At each clock rising edge, the PFD generates a control message according the the phase error between R and V, and it tries to bring R and V back in phase. During the next T_c time, the VCO creates some jitter due to the circuit noise which brings the next rising edge for V out of phase with R. This process repeats itself with a period of T_c. If the input is a clean clock with negligible jitter, each edge in the output clock should have the same amount of input-referred jitter, since they are identical edges. If we measure the input-referred jitter over a long time $\Delta T \gg 1/K_0$, the jitter at the beginning and the

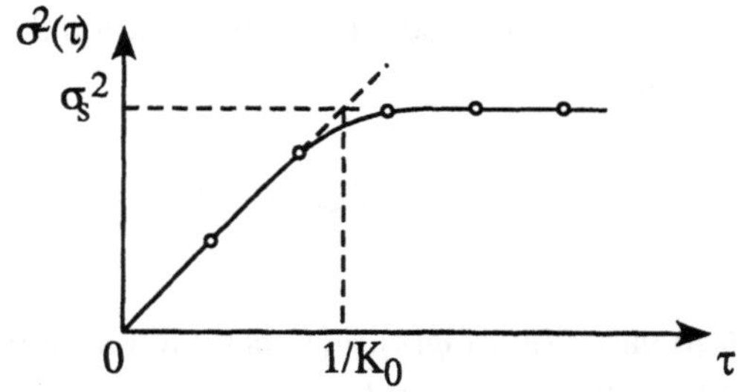

Figure 3.7. Self-referred jitter vs. time interval for a PLL

end of ΔT can be assumed to be uncorrelated due to such a long time interval. It can be seen from Figure 3.7 that the time interval jitter between the beginning and the end of ΔT is σ_s, the input-referred jitter of either of them has to be $\sigma_{abs}^2 = \sigma_s^2/2 = c/(2K_0)$.

The characteristics for both self-referred jitter and input referred jitter have been experimentally proved by McNeill in [McNeill, 1997]. Independently in our lab, it was observed to be true for wide bandwidth PLLs where the flicker noise for the VCO is suppressed by the negative feedback [Dai and Harjani, 2000a]. With flicker noise, the random walk process assumption in no longer true and the jitter between nearby transitions is partially correlated. In that case the self-referred jitter increases faster over the time intervals, since the jitter in the adjacent edges is partially in phase. We will not discuss jitter due to the flicker noise because of its mathematical complexity. However, the reduction of flicker noise for a CMOS ring VCO is our research focus and will be discuss in great details in later sections.

It is also worthwhile to point out that in practical PLL designs, there are usually several digital clocks running at different rates on the same chip. Hence cross talk is a common phenomenon. Then the jitter distribution is no longer Gaussian as is the case for white Gaussian noise. It is distributed at multiple peak time points instead of at a single average of kT_c. It is a result of mixing of multiple frequency components, and it is sometimes called *bimodal* or *multi-modal.*

3. Phase Noise vs. Timing Jitter

We have studied the concepts of both phase noise and timing jitter in the previous sections. Now we would like to discuss the relationship between them.

A periodic waveform can be either a sine wave, a square wave or of any other irregular but repetitive shapes. However, if we focus on their fundamental frequency component, they can all be written as Equation 3.6 with the presence of phase variation $\phi(t)$.

$$v(t) = A\sin[\omega_0 t + \phi(t)] \tag{3.6}$$

Due to the random variation of $\phi(t)$, $v(t)$ is no longer a periodic function of time. Timing jitter describes the amount of variation in its zero crossings relative to their ideal positions. It is a useful term for data communications where the precise control of the zero crossings is critical. The excessive variation of the zero crossings can cause increased BER. It is the magnitude of the time variation that is critical in data communications systems. It may contain the close-in phase components with a $1/f^3$ or $1/f^2$ shape as we have discussed before. It may also contain the energy of the frequency spurs which happens at the PFD update rate or any other frequencies due to the interference from other signals. It is the total power that has to be under control in order to achieve low clock jitter.

If we take the Fourier transform of Equation 3.6, we obtain its expression in the frequency domain as in Equation 3.7. For a noiseless waveform where $\phi(t) = 0$, Equation 3.7 becomes $\frac{A}{2j}[\delta(\omega - \omega_0) - \delta(\omega + \omega_0)]$, which means that it only has non-zero components at $\omega = \pm\omega_0$. Now the non-periodicity caused by $\phi(t)$ generates finite components in the vicinity of $\pm\omega_0$. The deviation in the frequency domain from the ideal spectrum is quantified as phase noise. It is an important term in wireless communications where the power spectrum has to be well under control in order to avoid cross talk between different communication channels. For instance, its close-in phase noise may pick up the signals in the adjacent channels, and a frequency spur at an offset frequency of $\Delta\omega$ may pick up the signals in the channels which is $\Delta\omega$ away. Therefore, the phase noise requirement is closely related to the frequency planning of the wireless systems. It is the power spectrum instead of the total noise power that is important in this case.

$$V(j\omega) = \int_{-\infty}^{+\infty} A\sin[\omega_0 t + \phi(t)]e^{-j\omega t}\,dt \tag{3.7}$$

Interestingly we can derive jitter from phase noise for some simple cases, or vice versa. For a stand-alone oscillator with white Gaussian noise, it can be shown that its self-referred jitter and phase noise can be written as in Equations 3.8 and 3.9, where τ is the time interval as we have discussed before [Demir et al., 2000].

$$\sigma^2(\tau) = c\tau \tag{3.8}$$

$$\begin{aligned} L(\Delta\omega) &= \frac{\omega_0^2 c}{\frac{1}{4}\omega_0^4 c^2 + \Delta\omega^2} \\ &\approx c\left(\frac{\omega_0}{\Delta\omega}\right)^2 \ (for\ |\Delta\omega| \ll \tfrac{1}{2}\omega_0^2 c) \end{aligned} \tag{3.9}$$

Please notice that the same factor c is used in both equations. This can be verified as follows. As we measure jitter with a time interval of τ, we are effectively observing phase noise at frequencies $\Delta f > 1/\tau$. If we integrate Equation 3.9 from $2\pi/\tau$ to ∞, we obtain Equation 3.10, which is the amount of jitter relative to the clock period. This is consistent with Equation 3.8.

$$\int_{\frac{1}{\tau}}^{\infty} L(\Delta\omega)\, d\Delta f = \int_{\frac{1}{\tau}}^{\infty} \frac{cf_0^2}{\Delta f^2}\, d\Delta f = -\frac{cf_0^2}{\Delta f}\bigg|_{1/\tau}^{\infty} = \frac{c\tau}{T^2} = \frac{\sigma^2(\tau)}{T^2} \tag{3.10}$$

The above derivation is only valid for white Gaussian noise. A more generic relationship for any wide sense stationary noise can be derived as follows. The self-referred jitter can be mathematically written as in Equation 3.11, where $E\{\}$ stands for the expectation value. We have assumed that $E[\phi^2(t)] = E[\phi^2(t+\tau)]$ because the noise is wide sense stationary.

$$\begin{aligned} \sigma^2(\tau) &= \frac{T^2}{(2\pi)^2} E\left\{[\phi(t+\tau) - \phi(t)]^2\right\} \\ &= \frac{1}{\omega_0^2} E\left[\phi^2(t) + \phi^2(t+\tau) - 2\phi(t)\phi(t+\tau)\right] \\ &= \frac{2}{\omega_0^2}\left\{E\left[\phi^2(t)\right] - E\left[\phi(t)\phi(t+\tau)\right]\right\} \end{aligned} \tag{3.11}$$

The auto-correlation function of $\phi(t)$ is defined as $R_\phi(\tau) = E\left[\phi(t)\phi(t+\tau)\right]$, and it is the Fourier transform of $S_\phi(\omega)$, the power spectral density of $\phi(t)$, i.e.,

$$R_\phi(\tau) = \frac{1}{2\pi}\int_{-\infty}^{\infty} S_\phi(\omega) e^{j\omega\tau}\, d\omega.$$

Therefore, Equation 3.11 can be further derived as in Equation 3.12, and it is a useful relationship between the self-referred jitter and phase noise, since $S_\phi(\omega)$ can often be replaced by the single-sideband phase noise $L(\omega)$.

$$\begin{aligned} \sigma^2(\tau) &= \frac{2}{\omega_0^2}\left[R_\phi(0) - R_\phi(\tau)\right] \\ &= \frac{1}{\pi\omega_0^2}\int_{-\infty}^{\infty} S_\phi(\omega)\left(1 - e^{j\omega\tau}\right) d\omega \\ &= \frac{2}{\pi\omega_0^2}\int_0^{\infty} S_\phi(\omega)\left(1 - \cos\omega\tau\right) d\omega \\ &= \frac{4}{\pi\omega_0^2}\int_0^{\infty} S_\phi(\omega)\sin^2\left(\frac{\omega\tau}{2}\right) d\omega \\ &= \frac{4}{\pi\omega_0^2}\int_0^{\infty} L(\Delta\omega)\sin^2\left(\frac{\Delta\omega\tau}{2}\right) d\Delta\omega \end{aligned} \tag{3.12}$$

In order to validate our derivation, let us take white noise as an example. Plugging Equation 3.9 into Equation 3.12, we get Equation 3.13, which is consistent with our expression for jitter.

$$\begin{aligned} \sigma^2(\tau) &= \frac{4c}{\pi} \int_0^{\infty} \frac{\sin^2\left(\frac{\Delta\omega\tau}{2}\right)}{\frac{1}{4}\omega_0^4 c^2 + \Delta\omega^2} d\Delta\omega \\ &= \frac{2}{\omega_0}\left(1 - e^{-\frac{1}{2}\omega_0^2 c\tau}\right) \\ &\approx c\tau \ (for\ c\tau \ll \frac{1}{\omega_0^2}) \end{aligned} \quad (3.13)$$

The condition of $c\tau \ll \frac{1}{\omega_0^2}$ is valid for all practical cases, since it means that the jitter is less than the clock period, hence the phase error does not roll over to the next period.

Similarly, if we take the approximation for Equation 3.9, i.e., $L(\Delta\omega) \approx \frac{c\omega_0^2}{\Delta\omega^2}$, and put it into Equation 3.12, we obtain $\sigma^2(\tau) = c\tau$. Thus it also validates Equation 3.12.

Even though Equation 3.5 describes self-referred jitter for white noise, it is invalid with the presence of flicker noise. Actually, the term of self-referred jitter is not well defined for flicker noise. Recall the jitter measurement setups in Figure 3.4. In order to measure the jitter variance, we have to wait long enough so that enough samples are collected. However, if significant amount of flicker noise exists, the amount of jitter grows as the measurement time elapses, since flicker noise has more components at low frequencies, which can only be observed over a long time. Now the self-referred jitter becomes a function of the observation time, and it has to be expressed as $\sigma(\tau, t)$, where t is the observation time. People generally wait for a period longer than than the time interval during which the signal is processed, or simply the system on-time. Therefore, the worst scenario is measured.

If the phase noise due to the flicker noise is expressed as $L(\Delta\omega) = \frac{\omega_0^2 K_f}{\Delta\omega^3}$, and the observation time is assumed to be t, the jitter can be derived as

$$\begin{aligned} \sigma^2(\tau, t) &= \frac{4}{\pi\omega_0^2} \int_{2\pi/t}^{\infty} \frac{\omega_0^2 K_f}{\Delta\omega^3} \sin^2\left(\frac{\Delta\omega\tau}{2}\right) d\Delta\omega \\ &= \frac{K_f \tau^2}{\pi} \gamma\left(\frac{t}{\tau}\right) \end{aligned} \quad (3.14)$$

where

$$\gamma\left(\frac{t}{\tau}\right) = \int_{\pi\tau/t}^{\infty} \frac{\sin^2 x}{x^3} dx \approx \ln(t/\tau) - 0.916 \ (for\ t/\tau \gg 1). \quad (3.15)$$

Even though integrating the phase noise from $1/\tau$ to ∞ as in Equation 3.10 gives the jitter $\sigma^2(\tau)$ for white noise, it is no longer valid for flicker noise. If we follow the same method, it does not predict the jitter as a function of observation time. The jitter predicted by this method is given by

$$\sigma^2(\tau) = \left(\frac{2\pi}{\omega_0}\right)^2 \frac{1}{2\pi} \int_{2\pi/\tau}^{\infty} \frac{\omega_0^2 K_f}{\Delta\omega^3}\, d\Delta\omega = \frac{K_f \tau^2}{2\pi} \tag{3.16}$$

Equation 3.16 clearly underestimates the jitter for flicker noise compared to Equation 3.14 in most practical cases.

We can also extend our discussion to include the calculation for the absolute jitter, where a reference clock is available at the PLL input. The absolute jitter can be derived as in Equation 3.17.

$$\begin{aligned}
\sigma_{abs}^2 &= \frac{1}{\omega_0^2} E\left[\phi^2(t)\right] \\
&= \frac{1}{\omega_0^2} R_\phi(0) \\
&= \frac{1}{\omega_0^2} \cdot \frac{1}{2\pi} \int_{-\infty}^{\infty} S_\phi(\omega)\, d\omega \\
&= \frac{1}{\pi\omega_0^2} \int_0^{\infty} S_\phi(\omega)\, d\omega \\
&= \frac{1}{\pi\omega_0^2} \int_0^{\infty} L(\Delta\omega)\, d\Delta\omega
\end{aligned} \tag{3.17}$$

We have derived the relationship between phase noise and timing jitter so far. Despite the different emphasis of phase noise and timing jitter, they are simply two different languages that describe the same phenomenon — the random variation in phase. Phase noise describes it in the frequency domain and timing jitter describes it in the time domain.

In this book, we study the methodology to minimize $\phi(t)$ in Equation 3.6. This results in the reduction in both phase noise and timing jitter. Therefore, we will not make a clear distinction between them and will use the term phase noise in most of the context to refer to both phase noise and timing jitter.

4. Conclusions

In this chapter we have reviewed the definitions for SSB phase noise and timing jitter. Based on the timing reference for the measurement, jitter can be categorized into input-referred jitter and self-referred jitter, and they have different characteristics. Phase noise and timing jitter are the same physical phenomenon described in two domains, namely frequency and time domains. This chapter serves as a foundation for further analysis in later chapters.

Chapter 4

REVIEW OF EXISTING VCO PHASE NOISE MODELS

In this chapter, we will first discuss the historical challenges of oscillator phase noise analysis, and then briefly go over a few outstanding works by D. B. Leeson, B. Razavi and A. Hajimiri. Finally we describe our research motivation that distinguishes our research from previous work.

1. Challenges in Oscillator Phase Noise Analysis

All oscillators are feedback systems. Consider the linear feedback system in Figure 4.1. The open-loop transfer function $H(j\omega)$ has to satisfy two conditions in Equation 4.1 at the frequency of oscillation ω_0. This is called Barkhausen's criteria. Positive feedback is formed under these conditions. Any signal at frequency ω_0 will sustain its initial amplitude and maintain oscillation.

$$\begin{cases} |H(j\omega_0)| = 1 \\ \angle H(j\omega_0) = 360^o \end{cases} \tag{4.1}$$

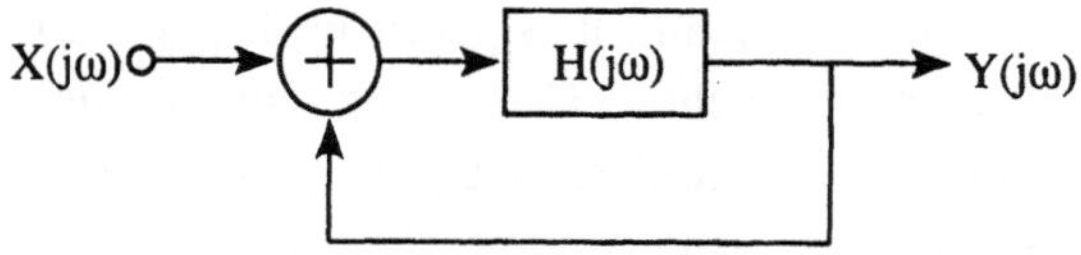

Figure 4.1. A linear feedback system

In practical designs, the oscillator has to start oscillation without any help of external stimuli. It has to amplify its internal noise or any external disturbance and keep increasing its amplitude until it reaches a satisfactory level. Therefore, in order to ensure the start of oscillation, $|H(j\omega_0)| > 1$ has to be satisfied at the start of the oscillation. As the amplitude of the signal increases, $|H(j\omega_0)|$ starts to drop as the circuit gradually saturates. Once the open-loop gain drops to one, the oscillation amplitude is maintained at a stable level. As a result, all oscillators, regardless of their types, are nonlinear systems. This is necessary to ensure the start of the circuit and maintain stable oscillation.

Furthermore, the oscillation amplitude is usually large enough to turn some of the circuit devices on and off periodically. Accordingly the transfer function $H(s)$ becomes time-variant.

Now that the oscillators are nonlinear time-variant systems, the traditional linear time-invariant (LTI) analysis is not valid for phase noise studies. Unlike linear amplifiers, in an oscillator, low frequency noise can be up-converted to the carrier vicinity and noise can also be aliased from the frequencies close to the harmonics of the carrier. This is the reason why the phase noise study of oscillators has been historically a difficult task.

Additionally, since the devices are turned on and off periodically, device noise, which contributes to phase noise, are modulated in the same manner. Therefore, it is no longer stationary. Its self-correlation function is now a periodic function of time. Hence it becomes cyclostationary instead.

In the rest of this chapter, we will review some of the existing phase noise models. Some are LTI models, and others are more complicated models. We will also briefly discuss their advantages and limitations.

2. Leeson's Model

Probably the most well-known phase noise model is Leeson's model which was proposed by D. B. Leeson in 1966 [Leeson, 1966]. He presented a heuristic derivation of the expected spectrum of a feedback oscillator in terms of known oscillator parameters without proof.

Leeson's model is expressed in Equation 4.2. $S_{\Delta\theta}(\Delta\omega)$ is the spectrum of the input phase noise uncertainty and it is given by Equation 4.3. $S_{\Delta\theta}(\Delta\omega)$ is expected to have two components. One component is the additive white noise at frequencies around the oscillator frequency, as well as noise at other frequencies mixed into the pass band of interest by nonlinearities. The second component is due to parameter variation at low frequencies around DC. The second component includes both white noise and flicker noise which has a power spectral density inversely proportional to frequency (a $1/f$ spectrum). This can be seen from Equation 4.3. In Equations 4.2 and 4.3, ω_0 is the center frequency of oscillation, Q is the loaded quality factor, $\Delta\omega$ is the offset frequency of interest, α is a constant determined by the flicker noise level, F is an empirical excess noise factor, k is the Boltzmann's constant, T is the absolute temperature, and P_s is the signal power.

$$L(\Delta\omega) = S_{\Delta\theta}(\Delta\omega)\left[1 + \left(\frac{\omega_0}{2Q\Delta\omega}\right)^2\right] \tag{4.2}$$

$$S_{\Delta\theta}(\Delta\omega) = \frac{\alpha}{\Delta\omega} + \frac{2FkT}{P_s} \tag{4.3}$$

Even though it seems intuitively true, Leeson's model was proposed without any formal proof. It considers the effect of the input noise in two regions as can be seen from Equation 4.2. When $\Delta\omega \ll \frac{\omega_0}{2Q}$, the input noise causes the same spectrum of the frequency variation with a multiplication factor of $\left(\frac{\omega_0}{2Q}\right)^2$. Hence the phase noise is multiplied by a factor of $\left(\frac{\omega_0}{2Q\Delta\omega}\right)^2$ where the $\Delta\omega$ in the denominator is because the phase is the integral of frequency. When $\Delta\omega \gg \frac{\omega_0}{2Q}$, the phase noise is the same of the input noise. In order to understand Leeson's model better, we will review two equivalent definitions of Q factor in this section.

Let us consider the RLC parallel network in Figure 4.2. Its conductance is calculated by Equation 4.4. We write its magnitude in Equation 4.5. At its resonant frequency $\omega_0 = 1/\sqrt{LC}$, the conductance becomes a real number and reaches its minimum value of $1/R$.

$$\begin{aligned} G(j\omega) &= \tfrac{1}{j\omega L} + j\omega C + \tfrac{1}{R} \\ &= \tfrac{1}{R} + j\left(\tfrac{1}{\omega L} + \omega C\right) \end{aligned} \tag{4.4}$$

$$|G(j\omega)| = \sqrt{\frac{1}{R^2} + \left(\frac{1}{\omega L} + \omega C\right)^2} \tag{4.5}$$

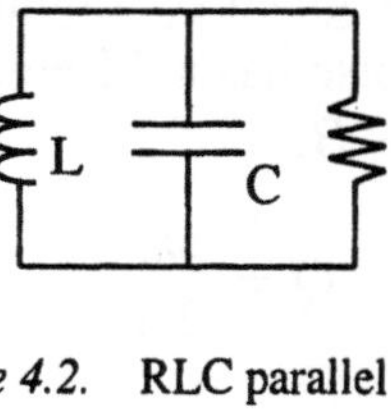

Figure 4.2. RLC parallel network

The most fundamental definition for Q factor is written in Equation 4.6. Assuming the RMS voltage value across the RLC tank is v_{RMS}, the expression for Q factor at the resonance can be derived as in Equation 4.7 where T is the period for the resonant frequency.

$$Q = 2\pi \frac{energy\ stored}{energy\ dissipated\ per\ cycle} \tag{4.6}$$

$$\begin{aligned} Q &= 2\pi \frac{v_{RMS}^2 C}{(v_{RMS}^2/R)T} \\ &= \omega_0 RC \\ &= \frac{RC}{\sqrt{LC}} \\ &= \sqrt{\frac{R^2 C}{L}} \end{aligned} \tag{4.7}$$

The magnitude of the impedance across the tank, which is the inverse of the quantity in Equation 4.4, is plotted in Figure 4.3. It reaches its maximum value of R at frequency ω_0. The frequency is $\omega_0 \pm B$ when the impedance drops to $R/\sqrt{2}$. An RLC network is a frequency selective system. $2B$ is its passband bandwidth around the center frequency of ω_0.

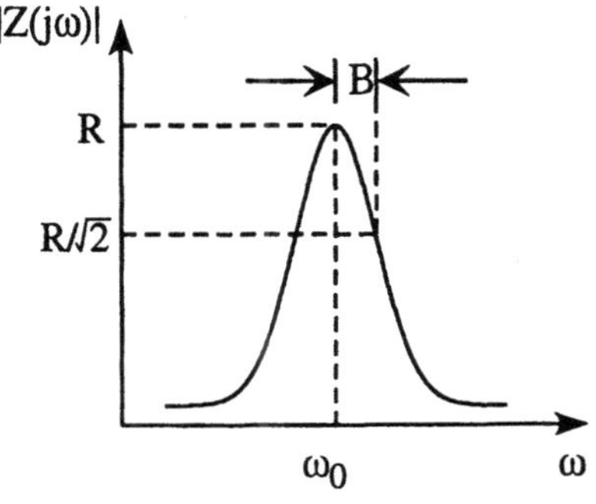

Figure 4.3. Magnitude of the impedance across the RLC tank

The second definition for Q factor is given by Equation 4.8. Equivalently, the bandwidth of the network is given by Equation 4.9. Recall that in Leeson's model, the output phase noise has a different response to the input noise depending on $\Delta\omega < B$ or $\Delta\omega > B$.

$$Q = \frac{\omega_0}{2B} \tag{4.8}$$

$$B = \frac{\omega_0}{2Q} \tag{4.9}$$

The two definitions of Q factor are equivalent. This can be seen from the following derivation. For small offset frequencies from ω_0, the conductance of the RLC network can be approximated as in Equation 4.10, and its magnitude is approximated in Equation 4.11. It is seen from Equation 4.11 that when $2\Delta\omega C = 1/R$, the magnitude of the impedance drops to $1/\sqrt{2}$ of its peak value. Hence the bandwidth B is given by $B = 1/(2RC)$. Now following the second definition, the Q factor can be derived as in Equation 4.12. This is the same value as we derived from the first definition as in Equation 4.7.

$$\begin{aligned} \frac{1}{Z[j(\omega_0+\Delta\omega)]} &\approx \frac{1}{Z(j\omega_0)} + \left[\frac{1}{Z(j\omega_0)}\right]' \Delta\omega \\ &= \frac{1}{R} + \left(jC - \frac{1}{j\omega_0^2 L}\right)\Delta\omega \\ &= \frac{1}{R} + \left(jC - \frac{C}{j}\right)\Delta\omega \\ &= \frac{1}{R} + 2j\Delta\omega C \end{aligned} \tag{4.10}$$

$$|Z(j\omega_0 + \Delta\omega)| \approx \frac{1}{\sqrt{(1/R)^2 + (2\Delta\omega C)^2}} \tag{4.11}$$

$$Q = \frac{\omega_0}{2B} = \frac{1}{\sqrt{LC}}\frac{1}{2/(2RC)} = \sqrt{\frac{R^2C}{L}} \tag{4.12}$$

From Equation 4.10, the conductance in the vicinity of the center frequency ω_0 can be expressed in terms of Q factor as in Equation 4.13.

$$\begin{aligned} \frac{1}{Z[j(\omega_0+\Delta\omega)]} &\approx \frac{1}{R}(1 + 2jRC\Delta\omega) \\ &= \frac{1}{R}\left(1 + 2j\sqrt{\frac{R^2C}{L}}\sqrt{LC}\Delta\omega\right) \\ &= \frac{1}{R}\left(1 + 2jQ\frac{\Delta\omega}{\omega_0}\right) \end{aligned} \tag{4.13}$$

In order to maintain steady oscillation, the lossy element R has to be compensated for by an active device which is modelled by a $-R$ in parallel with the LC tank. This is shown in Figure 4.4. Then the real part of $1/Z\left[j\left(\omega_0 + \Delta\omega\right)\right]$ in Equation 4.13 is cancelled out. Hence the impedance of the LC tank at resonance is approximated by Equation 4.14.

$$Z[j(\omega_0 + \Delta\omega)] = -jR \cdot \frac{\omega_0}{2Q\Delta\omega} \tag{4.14}$$

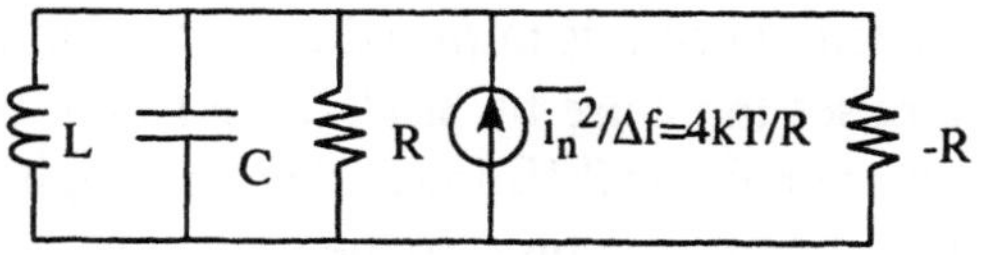

Figure 4.4. Model for an LC tank with an active device $-R$

Also shown in Figure 4.4 there is a noise current source associated with the resistor R, whose PSD given by $\overline{i_n^2}/\Delta f = 4kT/R$. Now the noise current flows through the LC tank and generates noise voltage v_n across it. Meanwhile, we assume that the RMS value of the signal voltage is $v_{s,RMS}$ and its average power dissipation within one cycle is $P_s = v_{s,RMS}^2/R$. Hence the noise-to-signal ratio is given by Equation 4.15.

$$\begin{aligned} \frac{\overline{v_n^2}/\Delta f}{v_{s,RMS}^2} &= \frac{(\overline{i_n^2}/\Delta f)\cdot|Z[j(\omega_0+\Delta\omega)]|^2}{v_{s,RMS}^2} \\ &= \frac{(4kT/R)R^2}{v_{s,RMS}^2}\left(\frac{\omega_0}{2Q\Delta\omega}\right)^2 \\ &= \frac{4kT}{v_{s,RMS}^2/R}\left(\frac{\omega_0}{2Q\Delta\omega}\right)^2 \\ &= \frac{4kT}{P_s}\left(\frac{\omega_0}{2Q\Delta\omega}\right)^2 \end{aligned} \tag{4.15}$$

The derivation of Equation 4.15 is based on a linear system assumption. Only the additive noise in the vicinity of ω_0 is considered. For modern communications systems, ω_0 is usually at a very high frequency. Hence for a CMOS oscillator, flicker noise is negligible at ω_0, and only thermal noise has to be considered.

We have derived the noise-to-signal ratio in the vicinity of ω_0. Now the question is how this quantity is related to phase noise. We can derive their relationship with MATLAB simulations.

We have simulated a signal at 1GHz with an additive noise at $1.4GHz$ which is 40dB below its level. The PSD plot is shown as the dashed curve in Figure 4.5. We then pass the signal and the noise through an amplitude limiter to remove the amplitude noise, assuming that there is no amplitude noise to phase noise conversion during the hard limiting. The resulting PSD is shown as the solid curve in Figure 4.5. The limiter has two effects on the noise spectrum. It lowers the original noise PSD at 1.4GHz by 6dB. Meanwhile, it mirrors the noise spectrum with respect to the carrier frequency.

If we consider white thermal noise in the vicinity of the frequency of oscillation, it has equal PSD on both sidebands. All oscillators inherently have an amplitude limiter internally. Hence the noise on each sideband is reduced by 6dB and mirrors itself to the other sideband. Since the noise on the two sidebands is uncorrelated, when it is mirrored and added to the other side, it is added in power and raises the noise by $3dB$. As a result, with the amplitude noise removed, the phase noise is $3dB$ lower than the original noise.

From Equation 4.15, the phase noise can be written as

$$L(\Delta\omega) = \frac{1}{2}\cdot\frac{\overline{v_n^2}/\Delta f}{v_{s,RMS}^2} = \frac{2kT}{P_s}\left(\frac{\omega_0}{2Q\Delta\omega}\right)^2 \tag{4.16}$$

In the above derivation, we have only considered the thermal noise from the resistor R. In practical designs, there is also noise from the active devices

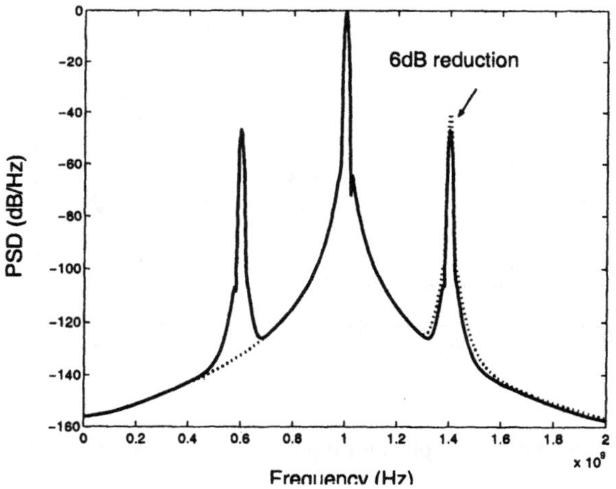

Figure 4.5. Additive noise to phase noise conversion

modelled by $(-R)$. Now that the circuit is nonlinear, there is noise aliased from DC and the harmonics of ω_0. Considering all these additional noise contributions, the close-in phase noise should be rewritten as

$$L(\Delta\omega) = \frac{2FkT}{P_s}\left(\frac{\omega_0}{2Q\Delta\omega}\right)^2 \tag{4.17}$$

where F is called the excess noise factor in Leeson's model. This is consistent with Leeson's model in Equation 4.2 for $\Delta\omega \ll \frac{\omega_0}{2Q}$.

Equation 4.16 indicates that there is a region for the offset frequencies where the phase noise PSD due to the thermal noise is proportional to $1/\Delta\omega^2$. This can be understood easily by the following argument. Within the bandwidth of B around the center frequency, the frequency selectivity of the LC tank is so weak that it can be assumed to be non-existent. Hence any noise could vary the frequency of oscillation. Since the phase is the integral of frequency over time, their relationship in the frequency domain is $\Phi(s) = \Omega(s)/s$. The PSD plots for the input device noise and the phase noise are illustrated in Figure 4.6. Within the region of $\omega_0/(2Q)$, due to the frequency modulation mechanism, the white thermal noise creates a phase noise slope of $1/\Delta\omega^2$, and the flicker noise creates a phase noise slope of $1/\Delta\omega^3$. Outside the bandwidth of $\omega_0/(2Q)$, the frequency selectivity of the LC tank prevents the oscillator from drifting. Hence any circuit noise could only cause instantaneous phase variation. Thus the phase noise follows the white spectrum of the thermal noise because of the phase modulation mechanism.

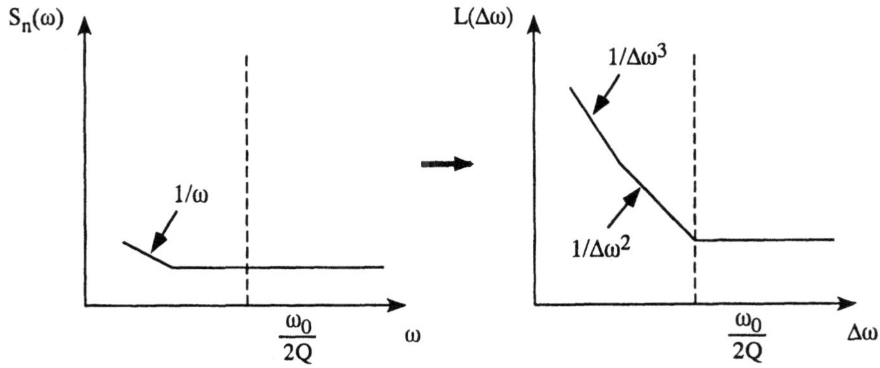

Figure 4.6. PSD plots for input device noise and phase noise

Leeson's phase noise model was proposed without proof. However, it has proved to be intuitively true. Its prediction of the phase noise spectrum, which includes $1/\Delta\omega^3$, $1/\Delta\omega^2$ and the white regions, is also verified by more stringent phase noise analysis developed later as well as numerous measurement results. It is written in a simple mathematical form. This makes it easy to use and understand. It includes an empirical factor F in its expression. Therefore it cannot predict phase noise from circuit noise analysis. However, once the phase noise for one oscillator is characterized, other oscillators of the same circuit topology can be calculated by applying the same F factor. Leeson's model assumes no knowledge of the circuit topology for the phase noise analysis. The topology dependency is taken into account by including the F factor. So it is unfortunate that Leeson's model does not provide a direction for circuit improvement. This disadvantage has been overcome by later phase noise theories.

3. Razavi's Model

Leeson's phase noise model is based on LC tank resonators. A monolithic CMOS design is preferred by modern communications systems. Unfortunately, the spiral inductors implemented in CMOS technologies take a large chip area, and they have low Q factors which are usually between 5 and 10. Oscillators based on LC tank have small frequency tuning range, and the desired frequency of operation may not reside within the limited tuning range over process and temperature variation. Therefore, inductorless VCOs have attracted tremendous research work in recent years. B. Razavi proposed a phase noise model in [Razavi, 1996] in 1996 for inductorless VCOs. In particular, his theory is well suited for CMOS ring oscillators.

Due to the lack of inductors, the definitions for the Q factor that we reviewed in the last section are no longer valid. In the case of an LC resonator, the energy is stored as electric energy in the capacitor and the magnetic energy in the inductor. Meanwhile part of the energy is dissipated in the resistor in parallel. Thus the definition for Q factor in Equation 4.6 is valid for LC tanks. In a ring oscillator, there are no inductors. The load capacitors are charged and discharged periodically. So there is no energy storage within a clock cycle. The definition for Q factor in Equation 4.8 does not apply to a ring oscillator either, since the closed-loop transfer function for a ring oscillator approaches infinity at the frequency of oscillator ω_0.

Razavi proposed a new definition for Q factor, which makes Leeson's model applicable to inductorless oscillators. If an oscillator is modelled as in Figure 4.1, and let $H(j\omega) = A(\omega)e^{j\Phi(\omega)}$, an open-loop Q factor is defined in reference [Razavi, 1996] as follows:

$$Q = \frac{\omega_0}{2}\sqrt{\left(\frac{dA}{d\omega}\right)^2 + \left(\frac{d\Phi}{d\omega}\right)^2} \tag{4.18}$$

We have reviewed Barkhausen's criteria at the beginning of this chapter. Both conditions in Equation 4.1 have to be satisfied in order to maintain steady oscillation. It is illustrated in Figure 4.7.

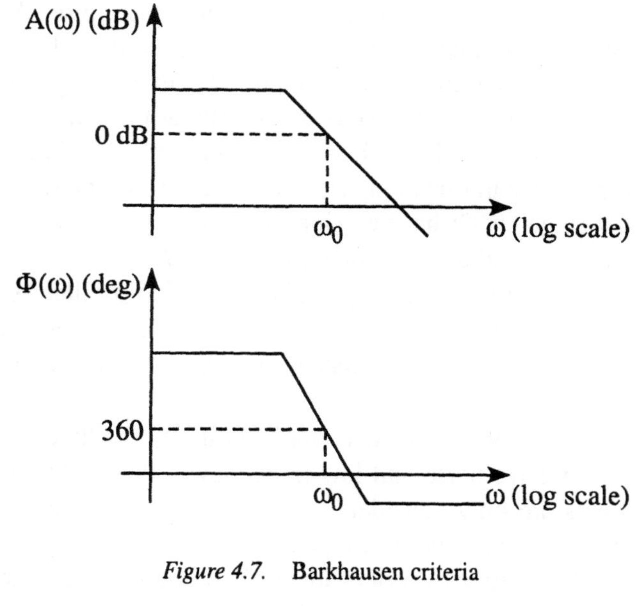

Figure 4.7. Barkhausen criteria

If there exists circuit noise that varies the transfer function instantaneously, the open-loop transfer function will deviate from Backhausen's criteria. The definition of Q factor in Equation 4.18 is a measure of how sensitive the open-loop transfer function is with respect to the circuit parameter variation. For example, an instantaneous increase in a transistor drain current due to noise means a larger transconductance g_m at that instance. Hence the open-loop transfer function $H(j\omega)$ is changed in both its magnitude and phase. A larger Q factor means more deviation from Backhausen's criteria. Therefore, for the same amount of circuit noise, there is a stronger feedback that brings the frequency back to its nominal frequency ω_0 so that both conditions in Backhausen's criteria are satisfied. It is interesting to note that this definition of Q factor is also consistent with the existing definitions of Q factor for LC resonators [Razavi, 1996].

Once the Q factor has been defined for a ring oscillator, it is implied in [Razavi, 1996] that Leeson's phase noise model in Equation 4.17 can be applied to the ring oscillator. However, in the case of an N-stage ring oscillator, there are N noise sources in total. So the close-in SSB phase noise for an N-stage ring oscillator is given by Equation 4.19.

$$L(\Delta\omega) = \frac{2NFkT}{P_s}\left(\frac{\omega_0}{2Q\Delta\omega}\right)^2 \tag{4.19}$$

If we maintain the linear assumption, the Q factor for a 3-stage ring oscillator is $3\sqrt{3}/4 \approx 1.3$, and the Q factor for a 4-stage ring oscillator is $\sqrt{2} \approx 1.4$. However, Q is not the only factor that determines the phase noise. A 4-stage ring oscillator has more noise sources than a 3-stage ring oscillator because of more delay stages. It also has to dissipate more power than a 3-stage ring oscillator with the same load capacitance in order to operate at the same frequency as a 3-stage ring oscillator does. For a fair comparison, only oscillators with equal power consumption should be compared. Considering all these factors, it can be shown that 4-stage VCOs have no significant advantage over 3-stage VCOs except for providing quadrature outputs [Razavi, 1996].

The same definition for Q can be applied to relaxation oscillators. It is shown in [Razavi, 1996] that the maximum Q for a relaxation oscillator is 1. Hence the phase noise for a relaxation oscillator is similar to, or worse than, that of a ring oscillator.

Reference [Razavi, 1996] considers nonlinearity by briefly discussing high-frequency multiplicative noise and low-frequency multiplicative noise. However, these effects are only considered as a minor modification to the linear model. In fact, this is only true when quasi-balanced delay stages are used. In ring oscillators where digital inverters are used, the signals are switched rail-to-rail. Figure 4.8 shows a 3-stage ring oscillator with digital inverter delay stages. Its internal voltages are also shown in the plot. One common misconception about ring oscillators is that it is close to its balanced state during the transition which is the most critical moment for phase noise, and hence a linear model is a good approximation to analyze its phase noise. However, it is seen from Figure 4.8 that the input and the output nodes of any of the delay stages never reach the balanced state together. Therefore, the delay stages rarely act as a linear amplifier even during the transition.

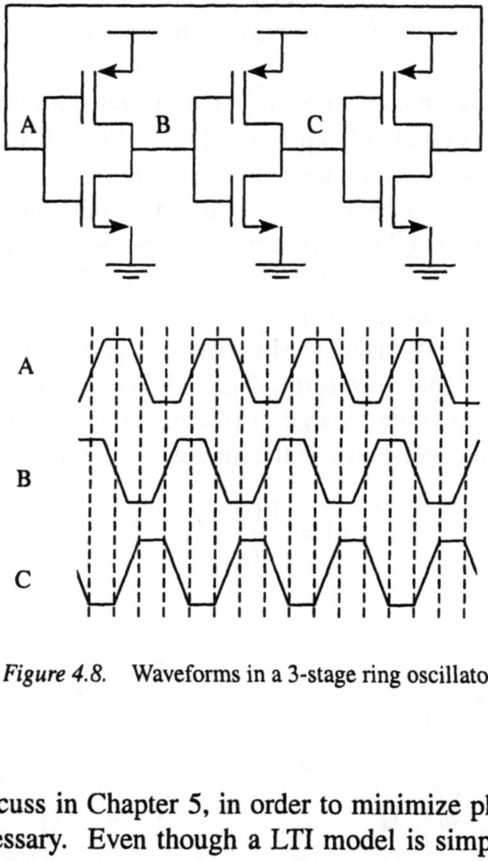

Figure 4.8. Waveforms in a 3-stage ring oscillator

As we will discuss in Chapter 5, in order to minimize phase noise, rail-to-rail swing is necessary. Even though a LTI model is simply and intuitive, a more accurate model is needed to consider the significant nonlinearity in a ring oscillator with large signal swings.

4. Hajimiri's Model

A more precise analysis was proposed by A. Hajimiri and T. Lee in 1998 [Hajimiri and Lee, 1998]. It introduces impulse sensitivity function (ISF or Γ) to consider the effects of nonlinearity, time-variance and cyclostationary noise.

Let us consider a current impulse injected into a circuit node. It changes the voltage waveform instantaneously in both amplitude and phase. Almost all the oscillators have an automatic amplitude control mechanism naturally built inside. Thus the amplitude will settle back to its original state after a certain time delay. However, a permanent phase shift will remain after it settles into its steady state. Phase noise stays forever once it is introduced into the circuit.

Hence phase noise is the accumulation, or integral, of circuit noise over time. This can be described as $1/s$ in frequency domain. This is consistent with our analysis in the previous section that the white noise causes $1/\Delta\omega^2$ phase noise.

The magnitude of phase shift is proportional to the ratio of the injected charge due to the noise to the maximum charge put on the capacitance on the node due to the signal swing, i.e., $\Delta q/q_{max}$. This holds reasonably valid for most of the noise sources.

On the other hand, phase noise depends on the time when the current is injected. This is illustrated in [Hajimiri and Lee, 1998] as shown in Figure 4.9. The circuit is most sensitive to noise during the transitions. An impulse at the zero crossing only causes phase noise, and it does not cause amplitude noise. This is shown in Figure 4.9 (a). An impulse at the peak only causes amplitude noise without any phase derivation. This is shown in Figure 4.9 (b). Hence the impulse response of phase is a step function whose amplitude depends on the time when the impulse is injected. Moreover, it is a periodic function with the same period as the signal waveform. The ISF is defined as in Equation 4.20 where q_{max} is the maximum charge displacement across the capacitor on the node, $u(t)$ is the unite step and τ is the time when the impulse is injected.

$$h_\phi(t,\tau) = \frac{\Gamma(\omega_0\tau)}{q_{max}} u(t-\tau) \tag{4.20}$$

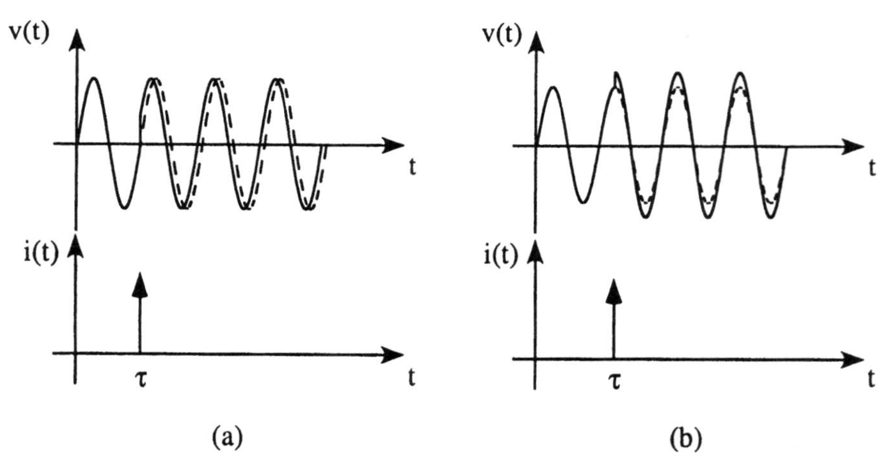

Figure 4.9. Perturbation of voltage waveform due to a current impulse at (a) the zero crossing (b) peak

It is worth mentioning that even though Figure 4.9 is intuitively true, it is inaccurate for most practical ring oscillator designs. The zero crossings are not the most noise sensitive points. Neither do the peaks have zero sensitivity. Figure 4.9 would have been valid only if there were no amplitude noise to phase noise conversion. Ideally the voltage limiting mechanism inside an oscillator removes the amplitude noise, and leave the phase noise unchanged. However, practical voltage limiters exhibit amplitude noise to phase noise conversion. Even though an impulse that occurs at the peak of the waveform only disturbs the amplitude instantaneously, part of the amplitude noise transfers into phase noise as it settles back to its original amplitude. A more accurate analysis [Demir et al., 2000] shows that ISF is still a periodic function, but it is shifted from what is described in [Hajimiri and Lee, 1998].

Nonetheless, the ISF provides an intuitive way of analyzing oscillator phase noise. It considers circuit nonlinearity, time variance and cyclostationary noise. Even though its assumption of the orthogonality of amplitude and phase does not hold exactly, it is a good approximation to avoid tedious computations in many cases. If a current impulse is injected into the circuit node in simulation, the ISF can be obtained by observing its phase shift after it settles to its steady state. Once ISF is obtained from circuit simulation, the rest of the theory holds and provides a clear picture of oscillator phase noise.

Given the ISF, the output excess phase $\phi(t)$ can be calculated using the superposition integral [Hajimiri and Lee, 1998]

$$\phi(t) = \int_{-\infty}^{\infty} h_\phi(t,\tau) i(\tau)\, d\tau = \frac{1}{q_{max}} \int_{-\infty}^{t} \Gamma(\omega_0 \tau) i(\tau)\, d\tau \tag{4.21}$$

Since the ISF is periodic function at frequency ω_0, only noise close to DC, ω_0 and its harmonics will result in non-zero excess phase as seen from the integral in Equation 4.21. Noise at all other frequencies will average out over time. The ISF can be expanded in a Fourier series as

$$\Gamma(\omega_0 \tau) = \frac{c_0}{2} + \sum_{n=1}^{\infty} c_n \cos(n\omega_0 \tau + \theta_n). \tag{4.22}$$

The coefficients c_i represent how much noise is contributed from the vicinity around frequency $i\omega_0$ where $i = 0, 1, 2, ...$ If the circuit has white noise with PSD of $\overline{i_n^2}/\Delta f$, its SSB phase noise is given by Equation 4.23.

$$L(\Delta\omega) = \frac{\Gamma_{rms}^2}{q_{max}^2} \cdot \frac{\overline{i_n^2}/\Delta f}{2\Delta\omega^2} \tag{4.23}$$

The SSB phase noise due to the flicker noise is given by

$$L(\Delta\omega) = \frac{c_0^2}{q_{max}^2} \cdot \frac{\overline{i_n^2}/\Delta f}{8\Delta\omega^2} \cdot \frac{\omega_{1/f}}{\Delta\omega} \tag{4.24}$$

Equations 4.23 and 4.24 indicate the $1/\Delta\omega^2$ region due to white noise and the $1/\Delta\omega^3$ region due to flicker noise. This is consistent with the conclusions in Leeson's model and Razavi's model.

Equation 4.24 shows that the phase noise due to flicker noise is proportional to c_0, the DC component for the ISF. This is a major contribution of this work. A common belief is that the phase noise $1/\Delta\omega^3$ corner is the same as the $1/f$ corner for flicker noise, since the complete spectrum close to DC is up-converted to the vicinity of the carrier. However, Equation 4.24 indicates that the $1/\Delta\omega^3$ region can be reduced by minimizing c_0. In an ideal case where the waveform is symmetrical, the $1/\Delta\omega^3$ is completely removed, since $c_0 = 0$.

Reference [Hajimiri and Lee, 1998] also considers cyclostationary noise by modulating $\Gamma(\omega_0\tau)$ by a factor $\alpha(\omega_0\tau)$ where $\alpha(\omega_0\tau)$ is the modulation factor due to the on and off operation of the devices. Unfortunately, the authors for [Hajimiri and Lee, 1998] did not distinguish the $\alpha(\omega_0\tau)$ for white thermal noise and the $\alpha(\omega_0\tau)$ for flicker noise. The PSDs for MOS device thermal and flicker noise are given by Equations 4.25 and 4.26 respectively.

$$\frac{\overline{i^2_{n,thermal}}}{\Delta f} = 4kT\gamma g_{ds0} \tag{4.25}$$

$$\frac{\overline{i^2_{n,flicker}}}{\Delta f} = \frac{K_f \cdot I_{ds}^{A_f}}{C_{ox}L_{eff}^2 f} \tag{4.26}$$

The γ factor in Equation 4.16 is thermal noise factor, and g_{ds0} is the output conductance of the transistor with zero V_{ds}. There exist several models for flicker noise, and Equation 4.26 is just one of them [Meta-Software, 1996]. K_f and A_f are both empirical numbers. It is seen from Equations 4.25 and 4.26 that the PSD for thermal noise is proportional to g_{ds0} while the PSD for flicker noise is proportional to $I_{ds}^{A_f}$. Therefore, the factor $\alpha(\omega\tau)$ is different for thermal noise and flicker noise. In order to minimize $1/\Delta\omega^3$ phase noise, it is crucial to minimize the DC component for $\alpha_{flicker}(\omega_0\tau)\Gamma(\omega\tau)$.

It is also worth mentioning that even though ISF is a good way of modelling phase noise, it has some practical difficulties. A current impulse as a δ function of time has to be injected into a circuit node in a simulation in order to obtain its phase response. However, only a current with finite amplitude and time duration can be simulated. A good approximation can only be achieved with a narrow impulse. Also a large instantaneous current can drive the circuit away from its normal operating state which is to be tested. Therefore, a current with small amplitude is also desirable. However, the current-time product defines the amount of charge injected. A small current amplitude and a short duration result in a small amount of charge injection. This is usually not sufficient to cause enough phase shift to be observed or subject to large numerical errors. This greatly limits the achievable accuracy of the ISF. In particular, there are noise insensitive nodes such as the power supply for a differential structure. In order to test the phase noise caused by the supply, the ISF for the power supply has to be simulated. However, this is very hard to do because of numerical errors.

Another practical drawback for ISF simulations is the long simulation time. In order to compute the ISF for a circuit node at time t, small time steps have to be taken to ensure accuracy and a long time is needed to allow the circuit to settle to its steady state after the impulse is injected. Therefore, the ISF at a single node and at a single time point needs a long transient simulation. In order to compute ISF over a complete period, many time points have to be taken, and this is only for a single circuit node. In order to evaluate the complete circuit, the ISF at each circuit node has to be simulated. As the circuit complexity grows, the complete computation for all the ISFs becomes so time-consuming that it eventually becomes impossible.

In conclusion, Hajimiri's model is a theory that provides a comprehensive phase noise analysis. Meanwhile it has some practical limitations. Compared with Razavi's theory, the ISF provides a more detailed analysis, however, it is highly simulation-based, and it does not trace the phase noise to its original sources. Hence it does not provide a direction of how to minimize phase noise by optimizing the circuit topologies.

In the next Chapter, we will propose our new phase noise model, which overcomes the shortcomings of the existing models. Our analysis is accompanied by simulation and measurement results that were used to verify it.

Chapter 5

UNIVERSAL MODEL FOR RING OSCILLATOR PHASE NOISE

In this chapter, we analyze the phase noise due to both the device noise and power supply/substrate noise. We approach our phase noise model by first providing some experimental comparison between five oscillators with different delay cell topologies in Section 1. Their oscillation frequency and the total power consumed are kept approximately equal for a fair phase noise comparison. Then we present a modified linear model in Section 2 which considers the nonlinear impact of voltage clipping. In Section 3, we define an effective Q factor (Q_{eff}) for the ring oscillators with voltage clipping and predict an increase in Q_{eff} with the advance in the operating speeds of CMOS technologies. In Section 4, we describe the noise up-conversion mechanism due to the bias and frequency control circuits. Then we discuss the impact of the digital switching noise coupled through the shared power supply and substrate in Section 5. Finally we provide some conclusions in Section 6.

1. Comparison and Analysis of Ring Oscillator Phase Noise

CMOS ring oscillators with five different delay cell topologies have been designed, fabricated and evaluated for phase noise performance. Our results in this section show that ring oscillators with linear loads provide much better phase noise performance than oscillators with nonlinear loads. We also observe that well designed single-ended oscillators have phase noise that is on par or better than oscillators with fully differential delay stages. Both our analysis and measurement results suggest that large signal voltage swing and improved linearity of the delay cells help reduce oscillator phase noise.

1.1 VCO Circuit Diagrams

In this subsection we introduce the five commonly used CMOS ring oscillators. Three of them are fully differential and two are single-ended. The circuit diagrams for the ring oscillators are shown in Figures 5.1 through 5.4. They are all three-stage CMOS voltage / current-controlled ring oscillators. However, only a single stage and its bias are shown in each of the figures. In Figure 5.1 the delay stage is implemented with a fully differential pair and combined PMOS loads (or called Maneatis loads, named after its inventer) [Maneatis, 1996]. It contains a source coupled differential pair and symmetric loads which provide good control over delay and high dynamic supply noise rejection. $M4-M7$ are equally sized and provide a more linear I-V characteristic than a single device load. The delay stage is biased by devices $M8$ - $M11$, and the delay time is controlled by the bias current I_b. Alternately, the load $M4 - M7$ of the delay stage can be implemented as a single PMOS device in either saturation or triode region as shown in Figure 5.2. Depending on the bias conditions of $M4$ and $M5$, they are either in the saturation or the triode region, which has an impact on phase noise as we will discuss in the following section. We also designed and fabricated the two single-ended ring oscillators shown in Figures 5.3 and 5.4. Figure 5.3 (a) is one of the three delay stages of a coupled ring oscillator. The transistor level schematics of the inverters and the transmission gate are drawn in (b) and (c). The circuit in (d) is used to generate the differential control voltages. Figure 5.4 shows a delay stage and its bias for a current-starved ring oscillator.

1.2 Phase Noise Analysis

In this subsection, we provide a qualitative discussion for the phase noise of the ring oscillators introduced in the previous section. In particular, the different types of loads in the delay cells are compared in terms of their impact on phase noise.

In most fully differential ring oscillators operating at high frequencies, the internal signal swings are limited to a fairly small range and the devices are rarely completely turned on or off. For most of the period the devices are close to their balanced states. As a result, the traditional linear model gives a reasonably good prediction for the phase noise. A possible exception is the ring oscillator with PMOS load in saturation, where the large load impedance results in a large internal voltage swing. The resulting non-linearity and cyclostationary noise make linear analysis less accurate. However, since the delay cells stay close to their balanced state for a significant portion of their period, we can, to first order, still expect linear analysis to provide insight into the phase noise performance.

We have reviewed Razavi's model in last chapter which predicts ring oscillator close-in phase noise to be give by Equation 5.1. However, Equation 5.1

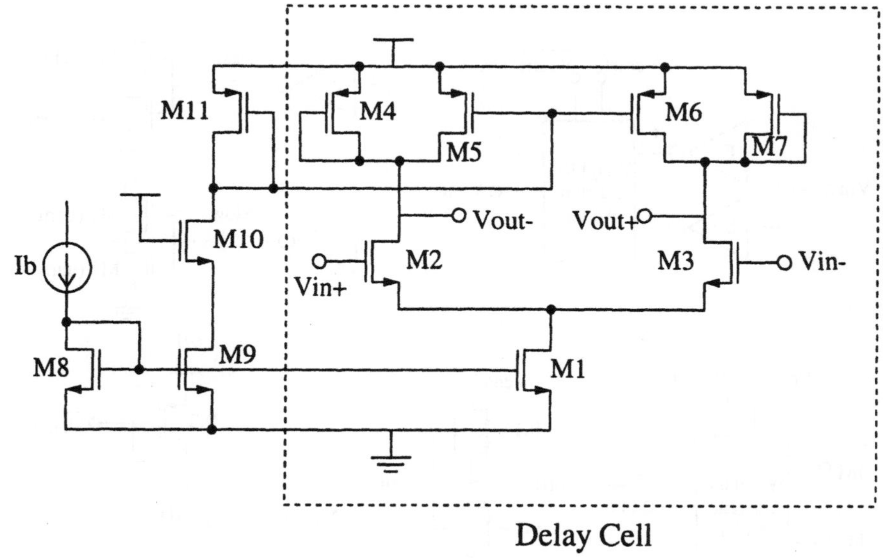

Figure 5.1. Differential inverter with Maneatis loads

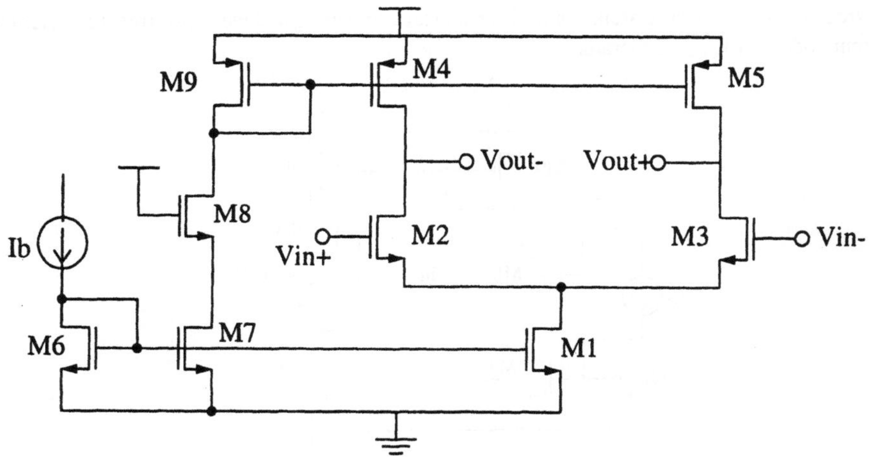

Figure 5.2. Differential inverter with single PMOS loads

is based on the assumption that the thermal noise is given by $\overline{v_n^2}/\Delta f = 4kTR$ where R is the parallel resistor in an RLC network. In our circuits, the thermal noise is given by $\overline{v_n^2}/\Delta f = 4kT\gamma g_{ds0}$ where g_{ds0} is the transistor output conductance g_{ds} for $V_{ds} = 0$. Here γ is the thermal noise factor, which is equal to 2/3 for long channel devices in saturation and can be as high as 2.5 for short channel devices [van der Ziel, 1986, Abidi, 1986, Klein, 1998].

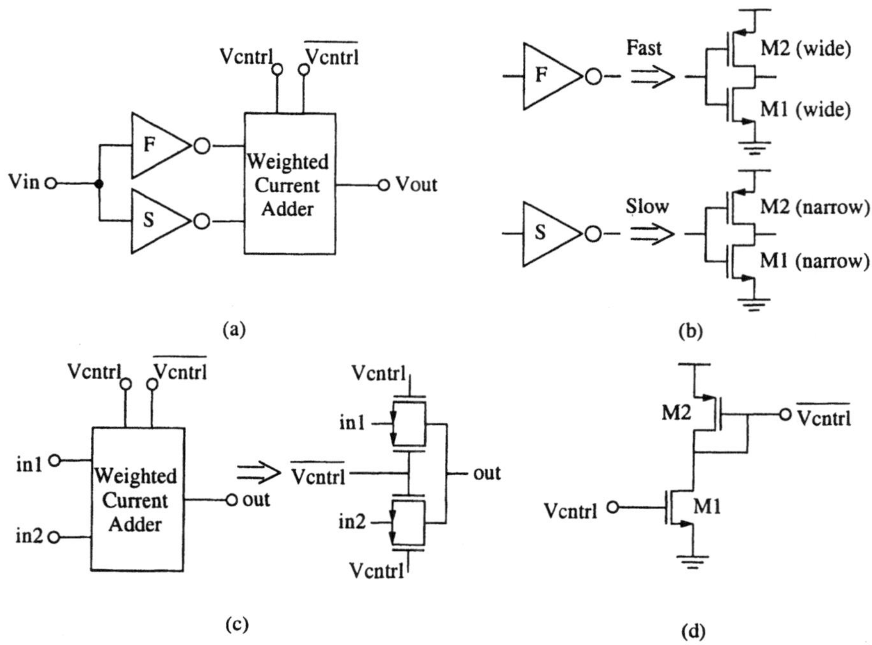

Figure 5.3. Coupled ring oscillator (a) inverter stage (b) strong and weak inverters (c) weighted current adder (d) $\overline{V_{cntrl}}$ generator

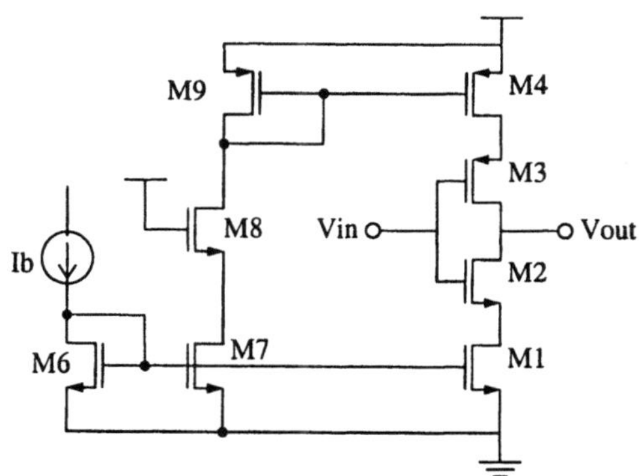

Figure 5.4. Current-starved inverter oscillator

$$L(\Delta\omega) = \frac{2NFkT}{P_s}\left(\frac{\omega_0}{2Q\Delta\omega}\right)^2 \tag{5.1}$$

Based on the above modifications and replacing Q with $3\sqrt{3}/4$ for three-stage ring oscillators, we obtain Equation 5.2.

$$L(\Delta\omega) = \frac{8F\gamma kT g_{ds0}}{9I_{rms}^2}\left(\frac{\omega_0}{\Delta\omega}\right)^2 \tag{5.2}$$

where I_{rms} is the RMS value for the internal current swing. It is a measure of the oscillation signal amplitude.

Since Q_L is independent of the delay cell topology, for a given center frequency ω_0 and offset frequency $\Delta\omega$, the phase noise is determined by F, I_{rms} and g_{ds0}. In summary, the oscillator phase noise is affected by two factors: the intrinsic device noise and the circuit noise sensitivity. F and g_{ds0} are factors that determine the device noise, and I_{rms} determines the circuit noise sensitivity. The circuit noise sensitivity in Equation(5.2) is consistent with the first-crossing theory [Abidi and Meyer, 1983] formula shown in Equation(5.3).

$$\sigma_T^2 = \frac{\overline{v_n^2}}{(dV/dt)^2} \tag{5.3}$$

where σ_T is the variation in the threshold-crossing time and v_n is the voltage noise. The first-crossing theory says that increasing the slew rate of the voltage helps reduce the timing jitter. In reality, this is achieved by increasing the current I_{rms}, which is consistent with our analysis.

It can be concluded from Equation (5.2) that in order to reduce phase noise, F and g_{ds0} have to be minimized and I_{rms} has to be maximized. For a given tail current and oscillation frequency, I_{rms} can only be increased by improved current switching efficiency, and this is only possible with increased voltage swing. In order to keep F and g_{ds0} low, the number of devices has to be minimized, the transistor widths cannot be too large, and $|V_{gs}|$ of the load devices has to be kept as small as possible to keep g_{ds0} low. Unfortunately, the design of these parameters are correlated. For example, efficient switching cannot be achieved with very narrow devices. Our measurement results provide additional insight into these design trade-offs.

Referring to Figure 5.2, when $M4$ and $M5$ are biased in triode, the equilibrium state of the output voltage swing is bounded by their gate overdrive voltage which is usually no more than a volt, and this results in a small voltage swing. On the other hand, when $M4$ and $M5$ are biased in saturation, large voltage switching is achieved due to the large output impedance, and this helps reduce the total phase noise.

Our analysis so far has assumed a linear system. However, in reality most oscillators are non-linear systems and cause noise aliasing [Hajimiri and Lee, 1998, Razavi, 1996]. The more nonlinear the system, the more likely that noise will get aliased from other frequencies. As a result, the circuit with a combined load in Figure 5.1 is expected to be a better choice than the circuit with a single load device in saturation in Figure 5.2. However, the improved linearity is achieved at the expense of reduced voltage swing.

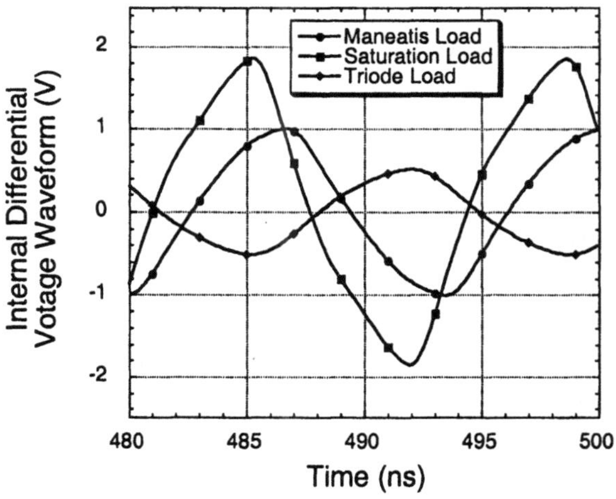

Figure 5.5. Internal differential voltage waveforms for ring oscillators with Maneatis load, saturation load and triode load

Figure 5.5 shows the internal voltage waveforms for the circuits in Figures 5.1 and 5.2. The ring oscillator with saturation load has the largest voltage swing and the fastest transitions, while the ring oscillator with triode load has the smallest voltage swing and the slowest transitions. On the other hand, the oscillator with saturation load has a more non-linear behavior than the other two as shown in Figure 5.5.

Intuitively a single triode device with a small V_{ds} is a relatively linear device. However, its voltage swing is limited by the gate overdrive voltage of the PMOS load. A large gate overdrive voltage is needed to allow a reasonable signal swing without driving the PMOS loads into saturation. Therefore, long and narrow devices with large gate overdrive voltage are preferred for the PMOS load. In our design, the PMOS loads are not sufficiently long and narrow such that in the balanced state, the output voltages stays close to the power supply voltage. Its swing is further hard limited by the supply rail. This results in significant non-linearity and small voltage swing.

In single-ended ring oscillators, most of the devices experience complete switching between on and off states, hence the simplified linear model for phase noise that has been used so far in this book does not apply. The large signal swing and the sharp transition improve the noise performance for the following reasons:

- Large signal swing implies large signal power and results in a large signal-to-noise ratio (SNR).

- The devices do not contribute additional noise when they are turned off.

- The short transition time results in a short period when the oscillator is sensitive to noise.

- The sharp transition edge minimizes noise sensitivity [Abidi and Meyer, 1983].

- Device flicker noise is reduced by the periodic on-off switching of the devices [Gierkink et al., 1999].

Despite the above benefits, it is a commonly held belief that single-ended circuits suffer from power supply and substrate noise injection when the oscillator is implemented on the same silicon substrate with digital circuits. As a result, fully differential oscillators have been widely used where low phase noise performance is critical. However, even with a fully differential implementation, the circuit is still subject to power supply and substrate noise due to the following reasons:

- Delay stages have poor common mode rejection when they are in an unbalanced state.

- The diffusion capacitances at the delay stage outputs are voltage-dependent. Therefore, they are modulated by the substrate and/or power supply noise.

- The threshold voltage of the input differential pair suffers from substrate noise because of back gate effects.

- The power supply and substrate noise causes additional charge injection into the common source of the differential pair through the parasitic capacitance, which results in frequency modulation and additional phase noise.

In a fully differential ring oscillator, the internal voltage swing rarely reaches rail-to-rail as in the case of the single-ended oscillator made of digital inverters. The small magnitude of the signal is more susceptible to any disturbance. As a result, single-ended ring oscillators are not necessarily worse than fully-differential ones in terms of substrate and supply rejection.

1.3 Measurement Results

The ring oscillators introduced in the previous subsections were designed and fabricated in an AMI $1.2\mu m$ CMOS process. The five different designs are: a differential oscillator with Maneatis loads, a differential oscillator with loads in triode, a differential oscillator with loads in saturation, a single-ended coupled-ring oscillator and a single-ended current-starved ring oscillator. We compare the phase noise measurement results for all the above designs in this subsection.

Figure 5.6. Die photograph of the test chip

Figure 5.6 is the die photograph of the test chip. There are five ring oscillators whose outputs are multiplexed and sent off chip through an output buffer. The output signal is measured with an HP E4408B spectrum analyzer and its SSB phase noise is calculated by using the following definition

$$L(\Delta\omega) = \frac{PSD\ @\ f = \omega_0 \pm \Delta\omega}{Total\ signal\ power} \tag{5.4}$$

All circuits were designed to operate with a $3.3V$ power supply and generate $80MHz$ output signal. However, due to process variations as well as inaccurate modelling and parameter extraction from the layout, they oscillate at slightly different frequencies and consume slightly different amounts of power.

In order to do a fair comparison, we have normalized the SSB phase noise by their center frequencies and their power consumption based on Equation 5.2. In order to compare the phase noise of the oscillators running at different frequencies, we use the relative offset frequency $\Delta\omega/\omega_0$ as the independent variable instead of the absolute offset frequency $\Delta\omega$. When the power consumption is scaled by a factor of α, it is equivalent to scaling all the device sizes by a factor of α, such that all the internal voltages remain the same and all the internal currents become α times as large. As a result, both I_{rms} and g_{ds0} are scaled by α. Referring to Equation (5.2), the phase noise is scaled by $1/\alpha$. Therefore, it is inversely proportional to the static power consumption of the oscillator. This validates our linear normalization.

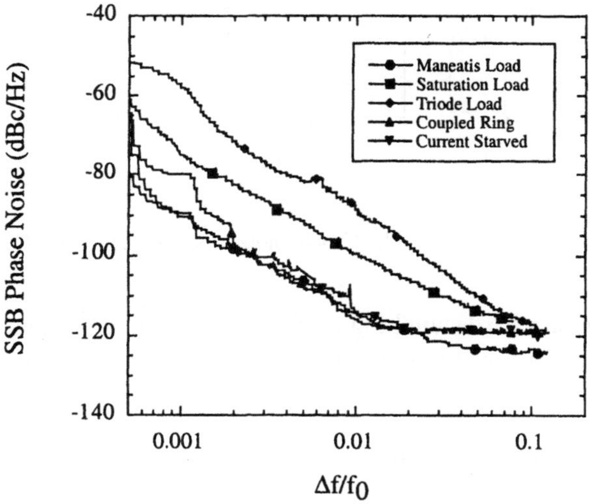

Figure 5.7. SSB phase noise of the ring oscillators

Figure 5.7 compares the measured SSB phase noise for the five different types of ring oscillators. They are plotted as a function of the relative offset frequency $\Delta f / f_0$ and scaled by the oscillator power consumption as discussed above. It can be seen that the fully differential ring oscillator with Maneatis load and the two single-ended ring oscillators have better phase noise performance than the fully differential ring oscillators with single device loads either in saturation or in triode. This confirms our previous conclusion that the linear characteristic provided by the Maneatis load should result in better phase noise performance. The oscillator with the triode load has the worst phase noise. This is not surprising, since it has the smallest internal signal swing. Additionally in our design, it suffers from non-linearity caused by the hard-limiting of the signal. The superior phase noise performance of the single-ended ring oscillators can also be understood by considering the benefits mentioned in the previous section about single-ended oscillators.

We have compared the phase noise for five types of ring oscillators in this section. Next we will introduce our theoretical model developed as a framework for ring oscillator phase noise analysis.

2. Modified Linear Model

In this section, we will derive the relation between the oscillator phase noise and the internal signal swing from a modified linear model for ring oscillators. This model applies to the differential pairs and the load devices in the delay cells of the ring oscillators. We show that, in order to achieve phase noise comparable with that of LC oscillators, fast transitions are needed, i.e., devices

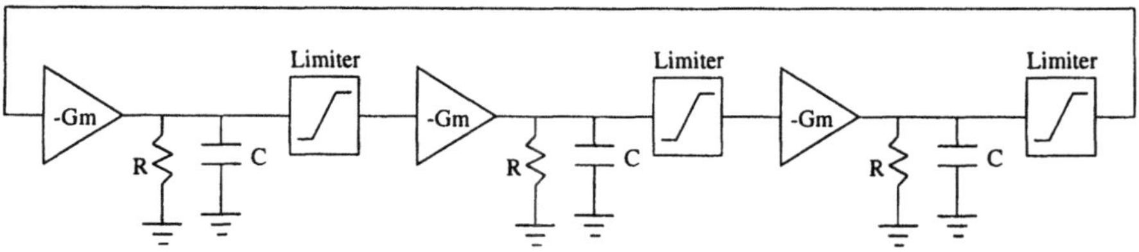

Figure 5.8. Modified linear model for a 3-stage ring VCO

have to operate in a hard switching mode and be switched on and off completely. Some measurement results for our test chips are also presented at the end of this section.

2.1 Theoretical Analysis

Figure 5.8 shows a simplified model for a 3-stage ring oscillator. Each delay stage consists of a resistor, a capacitor, a negative Gm cell and a voltage limiter. The system is linear as long as the internal voltages are not clipped by the limiters. When the peak-to-peak voltage swing tries to exceed the supply voltage of the circuit, the waveform becomes clipped by V_{dd} and ground, and this is modelled by the limiters in Figure 5.8. For this condition the Razavi's model [Razavi, 1996] for single-sideband (SSB) phase noise can be alternatively written as in Equation 5.5, where V_{pp} represents the peak-to-peak signal voltage, ω_0 is the center frequency of oscillation, and $\Delta\omega$ is the offset from the center frequency.

$$L\{\Delta\omega\} = \frac{64FkTR}{9V_{pp}^2}\left(\frac{\omega_0}{\Delta\omega}\right)^2 \tag{5.5}$$

Under linear operation, the voltage waveform is sinusoidal which can be written as $v(\omega_0 t) = \frac{V_{pp}}{2}\sin(\omega_0 t)$. For simplicity we have neglected its DC component and assumed that it is centered around $V_{dd}/2$. This is a good approximation for most optimally designed oscillators. Neglecting the DC term has no impact on our following analysis. It has been shown in [Hajimiri and Lee, 1998] that the *impulse sensitivity function* (ISF or Γ)[1] can be approximated as shown in Equation (5.6) while its RMS value is given by Equation (5.7).

$$\begin{aligned}\Gamma(\omega_0 t) &\approx \frac{dv(\omega_0 t)/d(\omega_0 t)}{|dv(\omega_0 t)/d(\omega_0 t)|_{max}^2} \\ &= \frac{2\cos(\omega_0 t)}{V_{pp}}\end{aligned} \tag{5.6}$$

$$\Gamma_{RMS}^2 = \frac{\omega_0}{2\pi}\int_0^{\frac{2\pi}{\omega_0}} |\Gamma(\omega_0 t)|^2\, dt = \frac{2}{V_{pp}^2} \tag{5.7}$$

Figure 5.9. Sinusoidal waveform clipped by power supplies

Comparing Equations (5.6) and (5.7), it is easy to see that the SSB phase noise for a three-stage ring oscillator can be expressed in terms of Γ_{RMS} as in Equation (5.8). Even though this expression is derived for a ring oscillator with linear operation, its validity extends into non-linear operation of a ring oscillator due to voltage clipping, since the ISF automatically considers the non-linear effects [Hajimiri and Lee, 1998].

$$L(\Delta\omega) = \frac{32FkTR\Gamma_{RMS}^2}{9}\left(\frac{\omega_0}{\Delta\omega}\right)^2 \tag{5.8}$$

Here, we have only considered thermal noise. Since the linear model does not predict noise aliasing, only noise close to ω_0 will cause phase noise in the above linear model. Even though flicker noise exists in CMOS oscillators, its magnitude is usually much smaller than the thermal noise at the oscillation frequency ω_0. Hence it is reasonable to neglect flicker noise in the linear model. A more rigorous analysis has shown that flicker noise does result in phase noise due to nonlinear and time-variant effects [Hajimiri and Lee, 1998]. Nevertheless, we will not discuss the impact of flicker noise here, since, as we will show in Section 4, the dominant flicker noise sources are the bias and frequency control circuits instead of the devices in the delay cell and it should be analyzed by a separate model.

In a real circuit, the waveform is bounded by the power supplies for large amplitudes. Let us assume that the sinusoidal waveform is symmetrically clipped as shown in Figure 5.9.

Its ISF can be approximated by Equation (5.9) where $k = 0, \pm1, \pm2, \ldots$.

$$\begin{aligned}\Gamma(\omega_0 t) &= \begin{cases} \frac{2}{V_{pp}}\cos(\omega_0 t) & (for\ \frac{V_{pp}}{2}\sin(\omega_0 t) < \frac{V_{dd}}{2}) \\ 0 & (for\ \frac{V_{pp}}{2}\sin(\omega_0 t) \geq \frac{V_{dd}}{2}) \end{cases} \\ &= \begin{cases} \frac{2}{V_{pp}}\cos(\omega_0 t) & (for\ k\pi - \arcsin(\frac{V_{dd}}{V_{pp}}) < \omega_0 t < k\pi + \arcsin(\frac{V_{dd}}{V_{pp}})) \\ 0 & (else) \end{cases}\end{aligned} \tag{5.9}$$

Figure 5.10. Waveform with soft slipping

The RMS value for the ISF can be derived as shown in Equation (5.10).

$$\begin{aligned}
\Gamma_{RMS}^2 &= \tfrac{\omega_0}{2\pi}\int_0^{\frac{2\pi}{\omega_0}} \Gamma^2(\omega_0 t)\,dt \\
&= \tfrac{\omega_0}{\pi}\int_{-\frac{1}{\omega_0}\arcsin(\frac{V_{dd}}{V_{pp}})}^{\frac{1}{\omega_0}\arcsin(\frac{V_{dd}}{V_{pp}})} \frac{4\cos^2(\omega_0 t)}{V_{pp}^2}\,dt \\
&= \tfrac{8}{\pi V_{pp}^2}\int_0^{\arcsin(\frac{V_{dd}}{V_{pp}})} \cos^2 x\,dx \\
&= \tfrac{4}{\pi V_{pp}^2}\int_0^{\arcsin(\frac{V_{dd}}{V_{pp}})} (1+\cos 2x)\,dx \\
&= \tfrac{4}{\pi V_{pp}^2}(x+\sin x\cos x)\Big|_0^{\arcsin(\frac{V_{dd}}{V_{pp}})} \\
&= \tfrac{4}{\pi V_{pp}^2}\left[\arcsin(\tfrac{V_{dd}}{V_{pp}}) + \tfrac{V_{dd}}{V_{pp}}\sqrt{1-\left(\tfrac{V_{dd}}{V_{pp}}\right)^2}\right]
\end{aligned} \tag{5.10}$$

We finally obtain the SSB phase noise expression given in Equation (5.11) by replacing Γ_{RMS}^2 in Equation (5.8) with the final expression in Equation (5.10)

$$\begin{aligned}
L\{\Delta\omega\} &= \tfrac{128FkTR}{9\pi V_{pp}^2}\left[\arcsin\left(\tfrac{V_{dd}}{V_{pp}}\right) + \tfrac{V_{dd}}{V_{pp}}\sqrt{1-\tfrac{V_{dd}^2}{V_{pp}^2}}\right]\left(\tfrac{\omega_0}{\Delta\omega}\right)^2 \\
&\approx \tfrac{256FkTRV_{dd}}{9\pi V_{pp}^3}\left(\tfrac{\omega_0}{\Delta\omega}\right)^2 (for\ V_{pp} \gg V_{dd})
\end{aligned} \tag{5.11}$$

We use V_{pp} in Equation (5.11) to represent the idealized peak-to-peak voltage of the sine wave as if there was no clipping. Equations (5.2) and (5.11) indicate that the phase noise is proportional to $1/V_{pp}^2$ for linear operation and is proportional to $1/V_{pp}^3$ when clipped by the power supplies. The additional noise reduction results from the fact that $\Gamma(\omega t) = 0$ when the voltage is clipped. Hence the period when the oscillator is susceptible to noise is reduced as the transitions take less time.

In reality, the clipping is rarely as hard as shown in Figure 5.9. We model this "soft clipping" shown in Figure 5.10 by Equation (5.12).

$$v(\omega_0 t) = \frac{V_{dd}}{2}\tanh\left[\frac{V_{pp}}{V_{dd}}\sin(\omega_0 t)\right] \tag{5.12}$$

The ISF is now approximated by Equation (5.13), and its RMS value is given by Equation (5.14).

$$\Gamma(\omega_0 t) = \frac{2\cos(\omega_0 t)}{V_{pp}\cosh^2\left[\frac{V_{pp}}{V_{dd}}\sin(\omega_0 t)\right]} \tag{5.13}$$

$$\begin{aligned} \Gamma^2_{RMS} &= \frac{1}{2\pi}\int_0^{2\pi}\left[\frac{2\cos x}{V_{pp}\cosh^2\left(\frac{V_{pp}}{V_{dd}}\sin x\right)}\right]^2 dx \\ &= \frac{2}{\pi V_{pp}^2}\int_0^{2\pi}\frac{\cos^2 x}{\cosh^4\left(\frac{V_{pp}}{V_{dd}}\sin x\right)} dx \end{aligned} \tag{5.14}$$

For $V_{pp} \ll V_{dd}$ Equation (5.14) can be approximated by Equation (5.15), which is consistent with the linear model.

$$\Gamma^2_{RMS} \approx \frac{2}{\pi V_{pp}^2}\int_0^{2\pi}\cos^2 x\, dx = \frac{2}{V_{pp}^2} \tag{5.15}$$

For $V_{pp} \gg V_{dd}$ the value of the function $\frac{\cos^2 x}{\cosh^4\left(\frac{V_{pp}}{V_{dd}}\sin x\right)}$ drops sharply as x increases from 0. Therefore, the integral in Equation (5.14) is dominated by the contributions for x values close to 0. In order to simplify the calculation, the approximation $\cos^2 x \approx \cos x$ can be made for $x \sim 0$. This has insignificant impact on the result in Equation (5.14) when $V_{pp} \gg V_{dd}$. Now Equation (5.14) can be approximated by Equation (5.16).

$$\begin{aligned} \Gamma^2_{RMS} &\approx \frac{2}{\pi V_{pp}^2}\int_0^{2\pi}\frac{\cos x}{\cosh^4\left(\frac{V_{pp}}{V_{dd}}\sin x\right)} dx \\ &= \frac{8}{\pi V_{pp}^2}\int_0^{\pi/2}\frac{\cos x}{\cosh^4\left(\frac{V_{pp}}{V_{dd}}\sin x\right)} dx \\ &= \frac{8}{\pi V_{pp}^2}\int_0^{\pi/2}\frac{d(\sin x)}{\cosh^4\left(\frac{V_{pp}}{V_{dd}}\sin x\right)} dx \\ &= \frac{8}{\pi V_{pp}^2}\int_0^{1}\frac{dx}{\cosh^4\left(\frac{V_{pp}}{V_{dd}}x\right)} \end{aligned} \tag{5.16}$$

Let $\alpha = \frac{V_{pp}}{V_{dd}}\sin x$. Then the result in Equation (5.16) can be further derived as shown in Equation (5.17).

$$\begin{aligned} \Gamma^2_{RMS} &= \frac{8V_{dd}}{\pi V_{pp}^3}\int_0^{\frac{V_{pp}}{V_{dd}}}\frac{d\alpha}{\cosh^4\alpha} \\ &= \frac{8V_{dd}}{\pi V_{pp}^3}\left(\frac{\sinh\alpha + 2\sinh\alpha\cosh^2\alpha}{3\cosh^3\alpha}\right)\Bigg|_0^{\frac{V_{pp}}{V_{dd}}} \\ &\approx \frac{8V_{dd}}{\pi V_{pp}^3}\cdot\frac{2}{3} \\ &= \frac{16V_{dd}}{3\pi V_{pp}^3} \end{aligned} \tag{5.17}$$

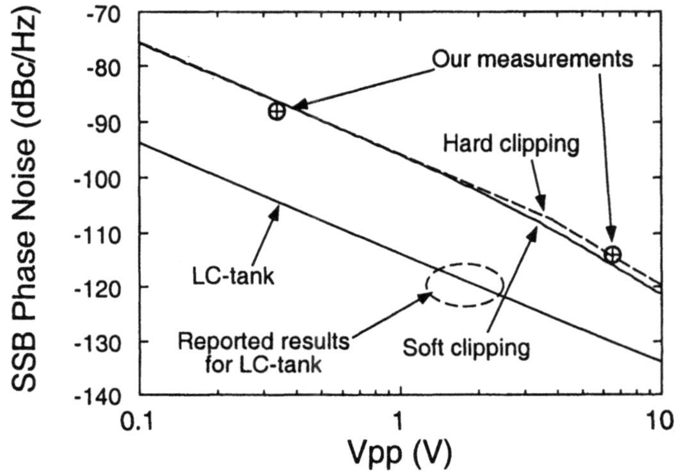

Figure 5.11. SSB phase noise vs. V_{pp}

Combining Equations(5.17) and (5.8), we can obtain the expression for the SSB phase noise as shown in Equation (5.18) with the soft clipping model.

$$L(\Delta\omega) = \begin{cases} \frac{64FkTR}{9V_{pp}^2}\left(\frac{\omega_0}{\Delta\omega}\right)^2 & (for\ V_{pp} \ll V_{dd}) \\ \frac{512FkTRV_{dd}}{27\pi V_{pp}^3}\left(\frac{\omega_0}{\Delta\omega}\right)^2 & (for\ V_{pp} \gg V_{dd}) \end{cases} \tag{5.18}$$

2.2 Simulation Results

In Figure 5.11 we show the SSB phase noise at 600 kHz offset from a 900 MHz carrier frequency as a function of different signal swings predicted by both of our hard clipping and soft clipping models. Here we assume that $F = 4, R = 1k\Omega$ and $V_{dd} = 3.3V$, which are typical values for practical designs.

Both results indicate that the phase noise consists of two regions with regard to V_{pp}: a $1/V_{pp}^2$ region without clipping and a $1/V_{pp}^3$ region with clipping. This implies the additional reduction in phase noise when the idealized voltage swing tends to exceed the power supply ($V_{pp} > V_{dd}$).

The two models provide slightly different break-even points between the two regions. The break-even point for the hard clipping model is $V_{pp} = 4V_{dd}/\pi \approx 1.27V_{dd}$, while the break-even point for the soft clipping model is $V_{pp} = 8V_{dd}/(3\pi) \approx 0.85V_{dd}$. The phase noise predicted by the two models are the same for small values of V_{pp} and only differ by 1.76dB for large values of V_{pp}. Since both models provide similar predictions for the phase noise, we expect the exact shape of the nonlinear limiting not to significantly impact the phase noise performance.

Since V_{pp} is directly related to the maximum slew rate by Equation (5.19) when the transitions take place, as an alternative, Equations (5.2), (5.11) and (5.18) can also be expressed in terms of $|dv/dt|_{max}$. This is useful when the waveform is clipped and V_{pp} is not directly available. We will not rewrite the above equations here, since the derivation is mathematically straightforward.

$$V_{pp} = \frac{2|dv/dt|_{max}}{\omega_0} \tag{5.19}$$

Maximizing V_{pp} is equivalent to maximizing the switching current. This suggests that improving the current switching efficiency can reduce the phase noise for a given supply current. For a ring oscillator with fully differential delay cells, an ideal case would be that all the tail current is used for switching, i.e., a current switching efficiency of 100%. In practical delay cell topologies, the maximal current for charging and discharging the load capacitors sets a lower bond on the phase noise for given power consumption.

A survey of published literature suggests that the phase noise of LC-tank oscillators is close to -120 dBc/Hz when the results are scaled to a 600 kHz offset from a 900 MHz center frequency [Craninckx and Steyaert, 1995, Craninckx and Steyaert, 1997, Kinget, 1998]. They are located within the ellipse in Figure 5.11. The phase noise curve for LC-tank oscillators is also plotted in this figure for $F = 4, R = 1k\Omega$ and $Q = 6$. Our calculation suggests that the phase noise for ring oscillators is likely to be much higher than that of LC-tank oscillators unless there is rail-to-rail switching, and efficient switching is the only possible approach.

2.3 Measurement Results

As a first exercise we have designed two types of ring oscillators to validate our phase noise model. Both of them have three delay stages. The bias circuit and the delay cell for one of them are shown in Figure 5.12 (called Maneatis ring oscillator from now on in this book). It contains a source coupled differential pair and symmetric loads which provide good control over delay and high dynamic supply noise rejection [Maneatis, 1996]. The other oscillator is a coupled ring oscillator whose diagram is shown in Figure 5.13. It contains two single-ended ring oscillators, namely a fast ring and a slow ring. Its frequency of oscillation can be tuned continuously between that of the fast ring and the slow ring.

Both oscillators were fabricated in an HP $0.5\mu m$ CMOS technology through MOSIS. the chip microphotographs are shown in Figure 5.14. As we are unable to measure the actual voltage swings without loading the oscillators, we

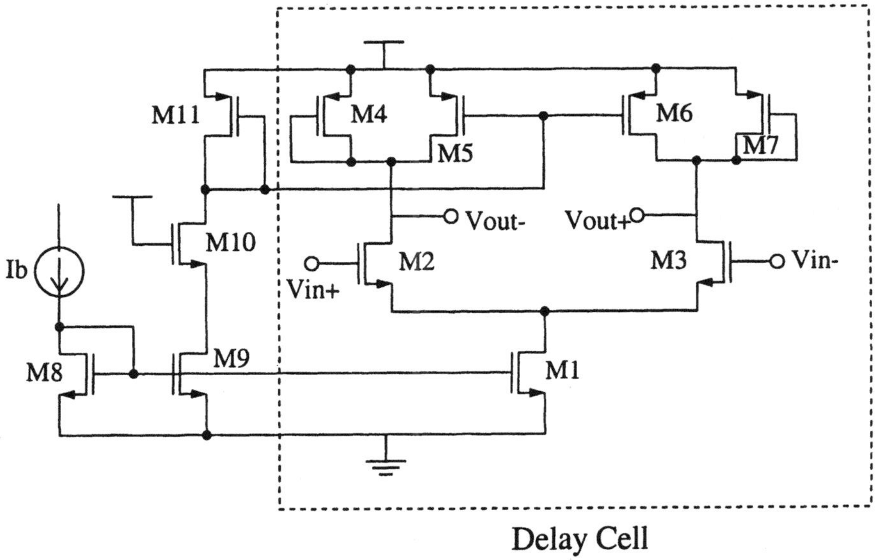

Figure 5.12. The bias and a delay cell for a ring oscillator with a source coupled pair and symmetric loads

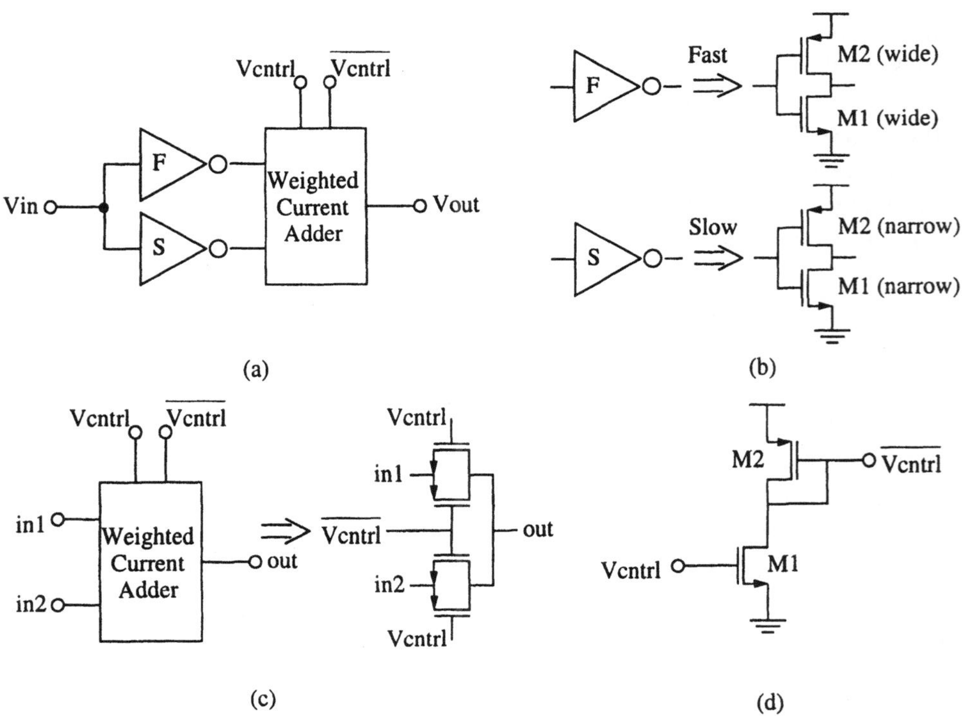

Figure 5.13. A three-stage coupled ring oscillator

Figure 5.14. Chip microphotographs for (a)Maneatis ring oscillator (b)coupled ring oscillator

use simulation results for V_{pp} as a guide. We derive V_{pp} from $|dv/dt|_{max}$ in the transient simulations for both oscillators. Our simulations suggest that $V_{pp} = 336mV$ for the Maneatis ring oscillator and $V_{pp} = 6.48V$ for the coupled ring oscillator. The measured center frequency for the Maneatis oscillator is 1.38GHz and the center frequency for the coupled ring oscillator is 960MHz. For a fair comparison, the phase noise for both oscillators are scaled to a 600kHz offset frequency from a 960MHz carrier. Our measured SSB phase noise results for the two oscillators are -88dBc/Hz and -114dBc/Hz respectively. Both measurement results are marked in Figure 5.11 with circled plus signs. They match our theoretical predictions very well for fairly typical values for F and R. The difference of 26dB/Hz in their SSB phase noise is primarily due to the difference in their values of V_{pp}.

In Figure 5.15 we plot the measured phase noise as a function of offset frequency. We note that the phase noise for the coupled ring oscillator is much lower than that for the Maneatis oscillator for all frequencies. It also rises faster at offset frequencies that are less than 300kHz. We have run transient circuit simulations and extracted the ISF's and operating conditions for every device in the coupled ring oscillator. The result suggests that the primary flicker noise contributors are the fast inverters. It could have been reduced further by a more symmetric waveform design [Hajimiri and Lee, 1998].

The simulation and measurement results provided in this section has validated our modified linear phase noise model. Our model suggests that the signal voltage swing is the most important factor that determines the oscillator phase noise due to the differential pairs and the loads in the delay cells. In essence, fast rail-to-rail switching is needed to minimize phase noise.

3. Q-factor for Ring Oscillators

In the last section we developed a modified linear model for the additive noise sources in a ring oscillator. In this section, we define an effective Q factor based on our model and predict its potential improvement for future technologies.

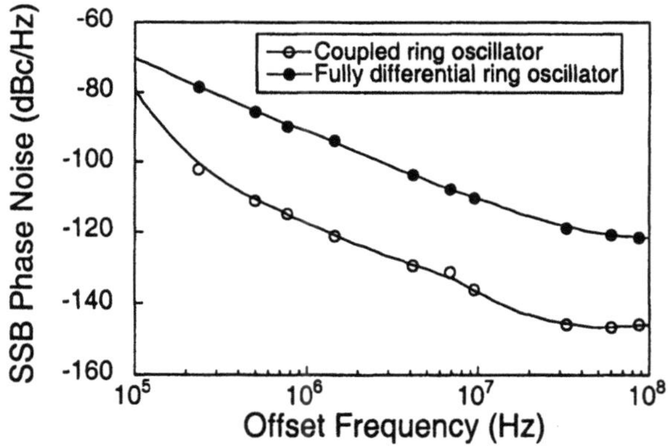

Figure 5.15. SSB phase noise comparison between the Maneatis oscillator and the coupled ring oscillator

For a linear model, the phase noise for an N-stage ring oscillator can be written in terms of the Q factor as [Razavi, 1996]

$$\begin{aligned} L(\Delta\omega) &= \frac{2NFkT}{P_s}\left(\frac{\omega_0}{2Q\Delta\omega}\right)^2 \\ &= \frac{4NFkTR}{V_{pp}^2}\left(\frac{\omega_0}{Q\Delta\omega}\right)^2 \end{aligned} \tag{5.20}$$

By comparing Equations (5.20) and (5.2) for $N = 3$, we can see that for a three-stage ring oscillator $Q = (3\sqrt{3})/4$ if no voltage clipping is considered.

In the last section we noted that voltage clipping helps reduce phase noise. Alternatively, we can think of this as an increase of the effective Q. By comparing Equations (5.20) and (5.18), Q_{eff} for a three-stage ring oscillator with large swings is given by Equation (5.21).

$$\begin{aligned} Q_{eff} &= \frac{9}{8}\sqrt{\frac{\pi V_{pp}}{2V_{dd}}} \\ &= \frac{9}{8}\sqrt{\frac{\pi |dv/dt|_{max}}{\omega_0 V_{dd}}} \end{aligned} \tag{5.21}$$

It is also interesting to express Q_{eff} in terms of the f_T of the CMOS process, which is given by $f_T = \frac{g_m}{2\pi C_g}$, where C_g is the total gate capacitance of the MOS device. The f_T characterizes the maximum speed for the devices in a certain technology[2]. It sets a lower bond for the transition time. Therefore, it is an important factor determining Q_{eff}.

When a digital inverter is used as a delay stage, the maximum slew rate is determined by the current flowing through the NMOS or PMOS device that charges or discharges the load capacitance. In order to minimize the flicker noise, the NMOS and the PMOS devices are sized to achieve identical rise and fall times [Hajimiri and Lee, 1998]. Hence, if $\mu_N = \gamma\mu_P$, where μ_N and μ_P

are the carrier mobilities for the NMOS and PMOS devices respectively, then the W/L ratio for the PMOS device has to be γ times larger than that for the NMOS device. Now that the waveform has the same rise and fall time, we can use the discharging process to calculate the transition time. Other than the gate capacitances of the next delay stage, the load usually contains some additional capacitance from an output buffer which amplifies the internal voltage and sends it to the output. The buffer helps prevent load pulling or the variation of frequency due to any change in the output load. Let us assume that the additional load capacitance due to the buffer is $\delta(C_{gN} + C_{gP})$, where C_{gN} and C_{gP} are the gate capacitances of the NMOS and PMOS transistors in the next stage respectively.

The maximum slew rate is now given by

$$\begin{aligned} \left|\frac{dv}{dt}\right|_{max} &= \frac{I_{d,max}}{(1+\delta)(C_{gN}+C_{gP})} \\ &= \frac{\frac{1}{2}g_{mN}(V_{dd}-V_{TN})}{(1+\delta)(C_{gN}+C_{gP})} \\ &= \frac{(V_{dd}-V_{TN})g_{mN}}{2(1+\delta)(1+\gamma)C_{gN}} \\ &= \frac{\pi f_{TN}(V_{dd}-V_{TN})}{(1+\delta)(1+\gamma)} \end{aligned} \tag{5.22}$$

Therefore, the Q_{eff} is now given by

$$Q_{eff} = \frac{9}{8}\sqrt{\frac{\pi f_{TN}}{2f_0(1+\delta)(1+\gamma)}\left(1-\frac{V_{TN}}{V_{dd}}\right)} \tag{5.23}$$

Next, we use Equation (5.23) to calculate the Q_{eff} as a function of the f_T for different supply voltages, assuming $\delta = 0$, i.e., negligible capacitance due to the buffer. For our analysis, we have assumed that $V_{TN} = 0.6V$. This assumption should hold reasonably well, since the device threshold voltage does not change substantially for different processes. The results are plotted in Figure 5.16. We have also calculated Q_{eff} for a number of different processes available via MOSIS, including TSMC $0.35\mu m$, TSMC $0.25\mu m$, TSMC $0.18\mu m$, HP $0.5\mu m$, HP $0.35\mu m$, AMI $1.5\mu m$, AMI $0.8\mu m$ and AMI $0.5\mu m$ [MOSIS,]. The f_T's for these processes were obtained by simulating a transistor with the gate voltage set to be equal to the power supply voltage. The maximum slew rates $|dv/dt|_{max}$ are derived from the measured results for ring oscillators from MOSIS. Then Equation (5.21) is used to calculate the effective Q. The results are summarized in Table 5.1 and also shown in Figure 5.16. It can be seen that there is a trend for higher Q_{eff} with larger f_T values for the various technologies. However it should be noted that the realizable Q_{eff}'s of future processes are not likely to increase as rapidly due to the reduced supply voltages and the lag in PMOS devices. Figure 5.16 projects that a Q_{eff} close to 4 is possible at 900 MHz for future processes. If this is achieved, the phase noise

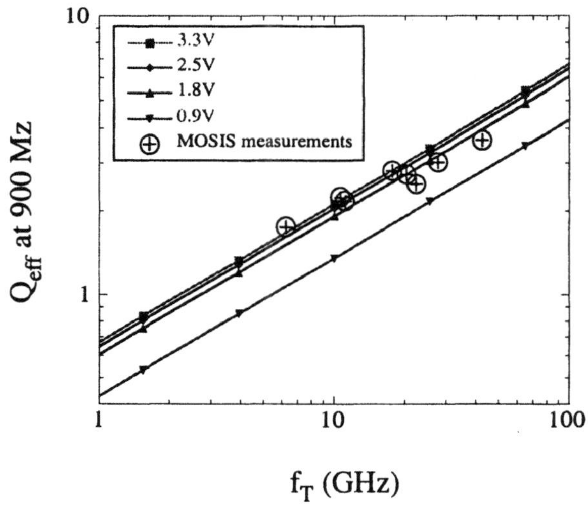

Figure 5.16. Effective Q factor as a function of the f_T of the process

PROCESS	f_T (GHz)	γ	$V_{TN}(V)$	$V_{dd}(V)$	Q_{eff}
TSMC 0.35μm	22.29	2.86	0.58	3.3	2.51
TSMC 0.25μm	27.65	4.77	0.52	2.5	3.02
TSMC 0.18μm	42.51	4.94	0.50	1.8	3.63
HP 0.5μm	17.64	2.96	0.66	3.3	2.80
HP 0.35μm	20.20	3.82	0.64	3.3	2.73
AMI 1.5μm	6.23	2.94	0.59	5.0	1.75
AMI 0.8μm	11.13	3.48	0.70	5.0	2.17
AMI 0.5μm	10.61	3.03	0.69	5.0	2.24

Table 5.1. Calculated Q_{eff} for processes available from MOSIS

for ring oscillators is likely to be comparable to that of LC-tank oscillators with on-chip spiral inductors. Furthermore, ring oscillators still retain their better integration properties and larger frequency tuning range.

4. Noise Up-Conversion

In a fully differential ring oscillator, low-frequency noise close to DC from the bias and the tail devices is up-converted to the vicinity of the carrier frequency by frequency modulation. This is not modelled by our analysis in the previous section or any previously published work. In this section, we analyze this noise up-conversion mechanism and validate it via simulations and measurement results.

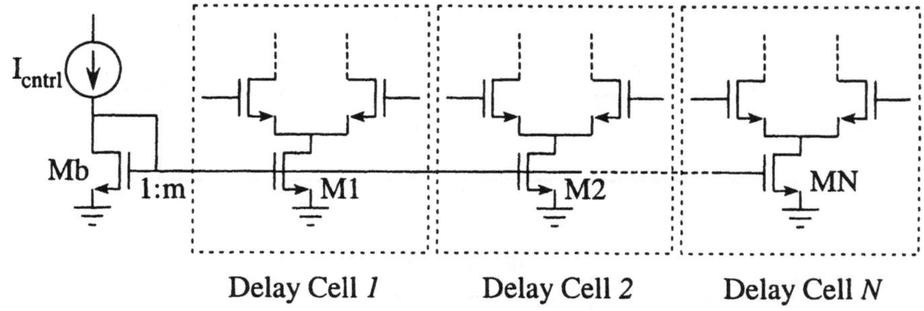

Figure 5.17. Bias structure for an N-stage ring oscillator

4.1 Theoretical Analysis

An N-stage ring oscillator is generally biased as shown in Figure 5.17. Noise in the tail current devices causes the g_m of the delay cell to change, hence varying the instantaneous oscillation frequency. We derive mathematical expressions to show the impact of bias noise on the overall oscillator phase noise. We will first consider thermal noise, and then extend our analysis to include flicker noise.

This up-conversion mechanism can be illustrated with the help of Figure 5.18. First, the noise is band-limited by the poles at the drains of M_b, M_1, M_2, ..., M_N. This low-pass filtered noise then modulates the g_m of the delay cells, and results in frequency variation. Finally, since $\phi(t) = \int_{-\infty}^{t} \omega(\tau)d\tau$ in the time domain, they satisfy $\Phi(s) = \Omega(s)/s$ in the frequency domain. This analysis suggests that the phase noise has a $1/f^2$ shape at low offset frequencies and drops off at a faster rate for higher offset frequencies. However, since the low-pass bandwidth is usually much larger than the oscillation frequency, the faster phase noise roll-off is rarely observed because other noise sources start to dominate. So it is of little interest. Additionally, we are usually more concerned with the up-conversion of the low-frequency noise. This is more of an issue when flicker noise is present and causes the phase noise to rise at a more rapid rate at low offset frequencies.

We treat the noise from the bias transistor M_b separately from the noise contributed by transistors M_1, M_2, ..., M_N in Figure 5.17, because the low-frequency noise from M_b is correlated for all stages while the noise from M_1, M_2, ..., M_N is not. In the following derivation, we assume that the current mirrors have a ratio of 1:m as shown in Figure 5.17.

In general, the oscillation frequency ω_0 is proportional to g_m/C_L, where C_L is the load capacitance for each delay cell. As an example, $\omega_0 = (\sqrt{3}/2)(g_m/C_L)$ for a 3-stage ring oscillator. Even though C_L includes a voltage-dependent part, its variation is usually much smaller than the variation in g_m in the presence of noise. We can therefore derive the relation in Equation (5.24) using this

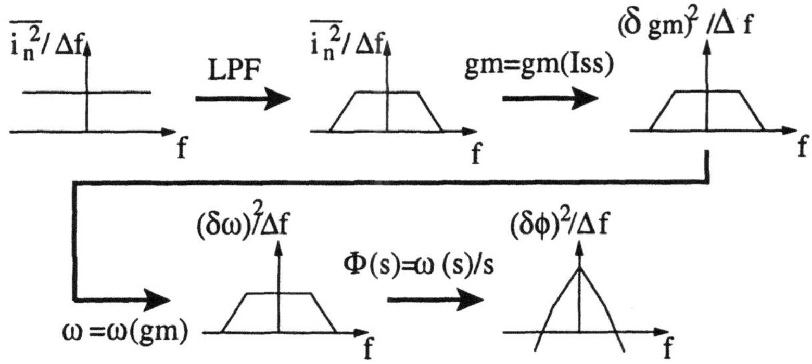

Figure 5.18. Up-conversion mechanism for low frequency noise in the tail and bias

assumption.

$$\left|\frac{\delta\omega}{\omega_0}\right| = \left|\frac{\delta g_m}{g_m}\right| \tag{5.24}$$

When operated in the long channel regime, the g_m of each delay cell is given by $g_m = \sqrt{(k'WI_{ss})/L}$, where I_{ss} is the tail current. This is valid for all reasonable gate overdrive voltages [Lee, 1998]. Using Equation (5.24), it can be shown that Equation (5.25) is valid for small signal amplitudes.

$$\left|\frac{\delta\omega}{\omega_0}\right| = \left|\frac{\delta g_m}{g_m}\right| = \frac{1}{2}\left|\frac{\delta I_{ss}}{I_{ss}}\right| \tag{5.25}$$

The thermal noise density from M_b is $\overline{i_n^2}/\Delta f = 4kT\gamma g_{ds0,b}$, where $g_{ds0,b}$ is the g_{ds0} for M_b. The noise power is amplified by m^2 times by each tail device. We can also replace $mg_{ds0,b}$ with g_{ds0} for the tail devices. Hence the normalized variation in frequency is given by Equation (5.26).

$$\frac{(\delta\omega)^2/\Delta f}{\omega_0^2} = \frac{(\delta g_m)^2/\Delta f}{g_m^2} = \frac{mkT\gamma g_{ds0}}{I_{ss}^2} \tag{5.26}$$

Finally we can write the resulting phase noise as shown in Equation (5.27).

$$\begin{aligned}\frac{(\delta\phi)^2}{\Delta f} &= \frac{1}{\Delta\omega^2}\cdot\frac{(\delta\omega)^2}{\Delta f}\\ &= \frac{\omega_0^2}{\Delta\omega^2}\cdot\frac{(\delta\omega)^2/\Delta f}{\omega_0^2}\\ &= \frac{mkT\gamma g_{ds0}}{I_{ss}^2}\left(\frac{\omega_0}{\Delta\omega}\right)^2\end{aligned} \tag{5.27}$$

The noise from M_1 to M_N are treated in a similar manner to the noise from M_b except that the noise is uncorrelated for each tail device. When we consider the frequency variation due to a single delay cell, Equation (5.25) has to be modified as shown in Equation (5.28) where $g_{m,i}$ and $I_{ss,i}$ are the g_m and I_{ss} for the ith delay cell.

$$\left|\frac{\delta\omega}{\omega_0}\right| = \frac{1}{N}\left|\frac{\delta g_{m,i}}{g_{m,i}}\right| = \frac{1}{2N}\left|\frac{\delta I_{ss,i}}{I_{ss,i}}\right| \tag{5.28}$$

Considering that there are N delay stages, the resulting phase noise expression is given by Equation (5.29).

$$\frac{(\delta\phi)^2}{\Delta f} = \frac{kT\gamma g_{ds0}}{NI_{ss}^2}\left(\frac{\omega_0}{\Delta\omega}\right)^2 \tag{5.29}$$

The SSB phase noise $L(\Delta\omega)$ is 1/4 of $(\delta\phi)^2/\Delta f$, since the variation in phase generates correlated phase noise on both sides of the carrier. Therefore, the total noise contributed by the bias transistor M_b (Equation (5.27)) and the N tail devices (Equation (5.29)) is given by Equation (5.30).

$$L(\Delta\omega) = \frac{1}{4}\frac{(\delta\phi)^2}{\Delta f} = \frac{\left(\frac{1}{N} + m\right)kT\gamma g_{ds0}}{4I_{ss}^2}\left(\frac{\omega_0}{\Delta\omega}\right)^2 \tag{5.30}$$

The up-conversion of flicker noise can be treated in a similar fashion. If the flicker noise is modelled as in Equation (5.31). [Meta-Software, 1996]

$$\frac{\overline{i_n^2}}{\Delta f} = \frac{K_f \cdot I_{ds}^{A_f}}{C_{ox}L_{eff}^2 f} \tag{5.31}$$

where K_f and A_f are flicker noise parameters. Then Equation (5.30) can be modified for flicker noise as shown in Equation (5.32).

$$L(\Delta\omega) = \frac{\pi(\frac{1}{N} + m)K_f \cdot \omega_0^2}{8C_{ox}L_{eff}^2 I_{ss}^{2-A_f}\Delta\omega^3} \tag{5.32}$$

Since m is usually much larger than $1/N$, the up-conversion of the noise from M_b is likely to be more severe than that from the tail devices. In practice, the flicker noise from M_b could be the major noise source at low offset frequencies.

To take into account of short channel effects, the drain-to-source current can be modelled as shown in Equation (5.33) [Foty, 1996].

$$I_D = \frac{\beta}{2}(1 - \epsilon_1)(V_{gs} - V_T)^{2-\epsilon_2} \tag{5.33}$$

where ϵ_1 and ϵ_2 are empirical numbers. They approach 0 for long channel devices and take on larger values for short channel devices. We use this format for short channel devices, because it maintains the simplicity of our equations. Here ϵ_1 represents the linear reduction in mobility due to short channel effects, and ϵ_2 represents the reduction of the exponent from the quadratic to the linear for extremely short devices. Values for ϵ_1 and ϵ_2 can be obtained from simulations or by equating Equation (5.33) with traditional short channel equations [Lee, 1998].

By following a similar derivation as above, it can be shown that

$$\left|\frac{\delta\omega}{\omega_0}\right| = \frac{1-\epsilon_2}{2-\epsilon_2}\left|\frac{\delta I_{ss}}{I_{ss}}\right| \tag{5.34}$$

The factor $(1-\epsilon_2)/(2-\epsilon_2)$ becomes smaller for short channel devices. As a result, both Equations (5.30) and (5.32) have to be multiplied by $4\left(\frac{1-\epsilon_2}{2-\epsilon_2}\right)^2$ to account for short channel effects.

4.2 Simulation Results

We have simulated a three-stage ring oscillator as an example whose bias circuit and one of the delay cells are shown in Figure 5.19. The frequency of the oscillator is controlled by the bias current I_{cntrl} supplied through M_6. We apply the analysis methodology proposed in [Hajimiri and Lee, 1998] by injecting a current pulse into the nodes 1, 2 and 3 throughout a complete clock cycle and observe the phase shift after it settles into steady state again. The effective ISFs (Γ_{eff}) which are modulated by the thermal and flicker noise are calculated for each individual device.

As mentioned in [Hajimiri and Lee, 1998], the RMS values for the Γ_{eff}s are important for thermal noise and the mean values of the Γ_{eff}s are important for flicker noise. Table 5.2 lists the RMS values for $\Gamma_{eff,thermal}$s and mean values for $\Gamma_{eff,flicker}$s for all the transistors. It can be seen that the thermal noise contribution of all the devices are similar in magnitude, though the contribution of $M_{2,3}$ is slightly larger than the others. However, for flicker noise the contributions of M_1 and M_6 dominate, suggesting that the majority of the low frequency phase noise is contributed by the bias devices.

Table 5.2. $\Gamma_{RMS,eff,thermal}$ and $\overline{\Gamma_{eff,flicker}}$

RMS/Mean for Γ	M_1	$M_{2,3}$	$M_{4,5}$	M_6
Thermal ($deg/sec^{1/2}$)	25	89	55	42
Flicker (deg/sec)	-2128	35	-168	7231

We have only included a single device M_6 in the bias. In reality, more devices are often used for the bias. For example, a voltage-to-current converter

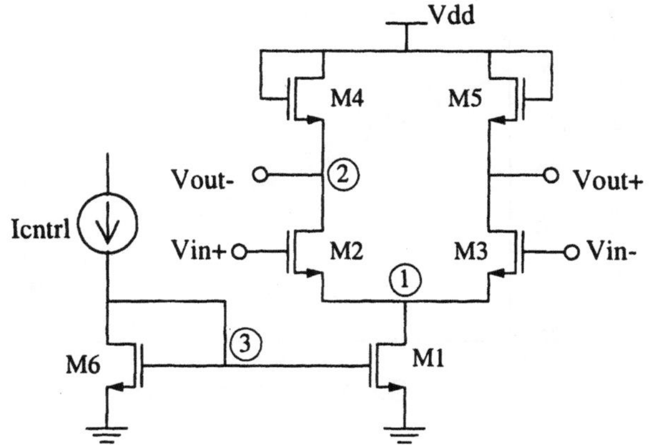

Figure 5.19. The bias and a delay cell for a three-stage ring oscillator

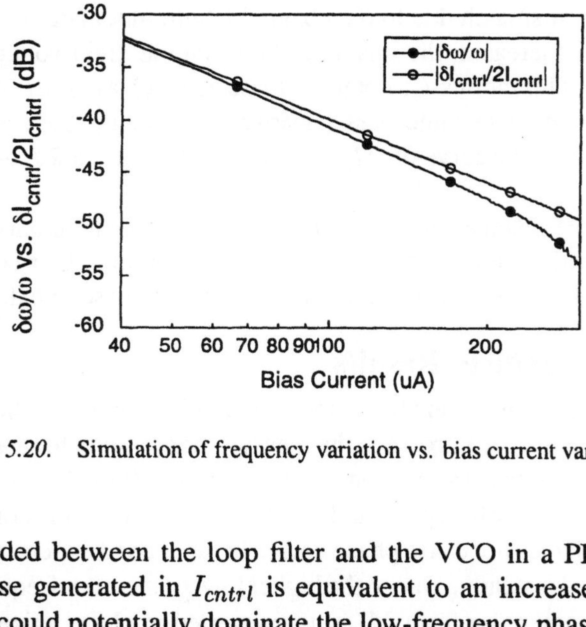

Figure 5.20. Simulation of frequency variation vs. bias current variation

is usually needed between the loop filter and the VCO in a PLL. Any low-frequency noise generated in I_{cntrl} is equivalent to an increase in the noise from M_6 and could potentially dominate the low-frequency phase noise.

We verify Equation (5.25) in our noise up-conversion model by simulating the circuit in Figure 5.19. We vary the bias current, and observe the variation in frequency in the steady state. We assume I_{ss} is proportional to I_{cntrl} and $\partial\omega/\partial I_{ss}$ is the same for DC and any low-frequency variation in I_{ss}. Figure 5.20 provides the comparison between $|\delta\omega/\omega|$ and $|\delta I_{cntrl}/2I_{cntrl}|$.

In Figure 5.20 both $|\delta\omega/\omega|$ and $|\delta I_{cntrl}/2I_{cntrl}|$ are drawn as functions of the bias current. In the simulations δI_{cntrl} is given a fixed value so that I_{cntrl} is swept linearly. Therefore, the curve of $|\delta I_{cntrl}/2I_{cntrl}|$ drops at a slope of 20 dB/decade. The two curves match extremely well for low bias currents. The

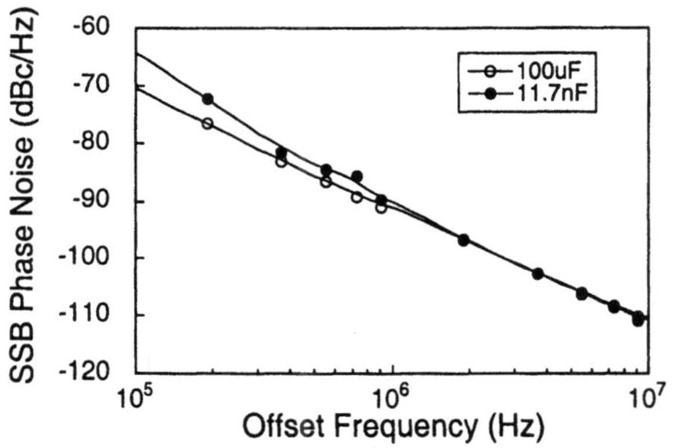

Figure 5.21. Phase noise with different by-pass capacitors

error increases with larger bias current because of short channel effects and current mirror mismatch due to the channel length modulation.

When I_{cntrl} increases, the mismatch between the drain voltages of M_6 and M_1 increases. This causes I_{ss} not to increase as much as I_{cntrl}. As a result, the oscillation frequency becomes less sensitive to I_{cntrl}, and the curve of $|\delta\omega/\omega_0|$ falls further below the curve of $|\delta I_{cntrl}/2I_{cntrl}|$ in Figure 5.20 for the increased bias currents.

Despite short channel effects and channel length modulation effects, our simulations confirm our model for noise up-conversion. Our conclusions are further strengthened by our measurement results discussed next.

4.3 Measurement Results

Again, we use measurement results for the Maneatis oscillator shown in Figure 5.12. Figure 5.21 provides the measurement results for the circuit tested under two conditions. In the first experiment, a $100\mu F$ by-pass capacitor is connected between the bias point and ground so that most of the noise from M_6 is filtered out and does not cause phase noise. In the second experiment, a $11.7nF$ capacitor is used instead[3]. Due to the higher low-pass corner frequency, some of the low-frequency noise from M_6 is up-converted to phase noise by frequency modulation. With a large by-pass capacitor, the phase noise curve retains the $1/f^2$ characteristic of the thermal noise. With a small by-pass capacitor, the phase noise curve starts to rise faster at low offset frequencies and enters the $1/f^3$ region. This implies that the flicker noise from the bias transistor can dominate at low frequencies which confirms our previous theoretical analysis.

In a practical PLL design, the bias point is an internal node and no large off-chip capacitor is available to by-pass the flicker noise from M_6 and other transistors in the bias circuit. Furthermore, no additional low-frequency poles can be placed at the VCO frequency control port in a PLL due to stability con-

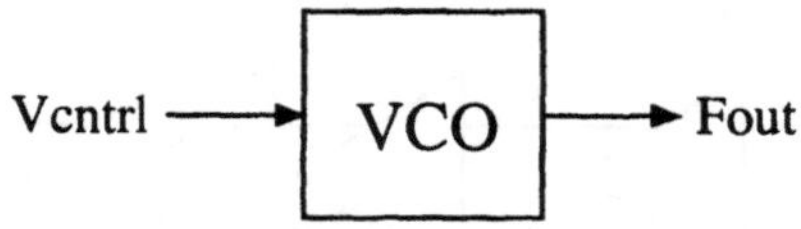

(a)

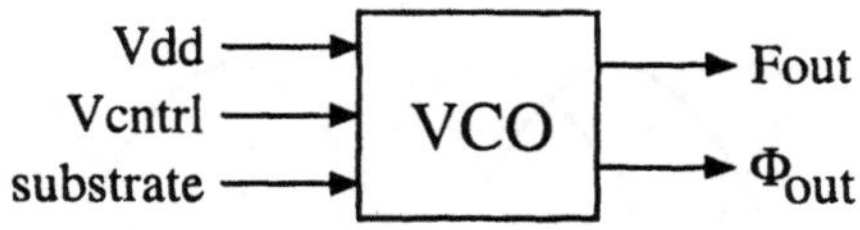

(b)

Figure 5.22. A VCO block diagram (a) traditional view (b) a rigorous view

cerns. Therefore, without this filtering possibility the flicker noise level can become so high that it is even dominant at frequencies beyond the loop bandwidth and cannot be suppressed by the PLL. This is particularly an important issue for low-bandwidth PLLs. Therefore, for integrated VCO designs, the current mirror bias structure should be avoided if possible.

5. Power Supply / Substrate Noise

So far we have analyzed the impact on phase noise caused by intrinsic device noise. When the oscillator is fabricated on the same silicon substrate and shares the same power supply with digital circuits, there is a significant amount of noise due to the digital switching activity coupled through the common power supply and the substrate. In this section, we will analyze the impact of the supply and substrate noise.

5.1 Theoretical Analysis

The output frequency of the VCO is traditionally considered only as a function of the control voltage, and the VCO is viewed as a block which does frequency modulation (FM). It has one input port and one output port. This is shown in Figure 5.22 (a). However, a more rigorous view of the VCO is shown in Figure 5.22 (b). The VCO is not only controlled by V_{cntrl}, but also the power supply and the substrate. In addition to frequency variations, changes in the inputs also cause instantaneous phase shift at the output, resulting in phase modulation (PM) as well. Since a change in the output frequency causes the phase noise to accumulate, FM is usually the dominant factor, and PM can be neglected. In the following discussion, we will focus on the power supply noise. The impact of the substrate noise can be derived similarly.

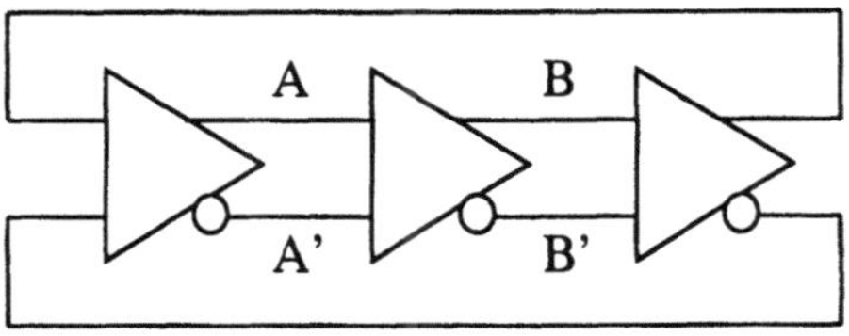

Figure 5.23. A three-stage differential ring oscillator

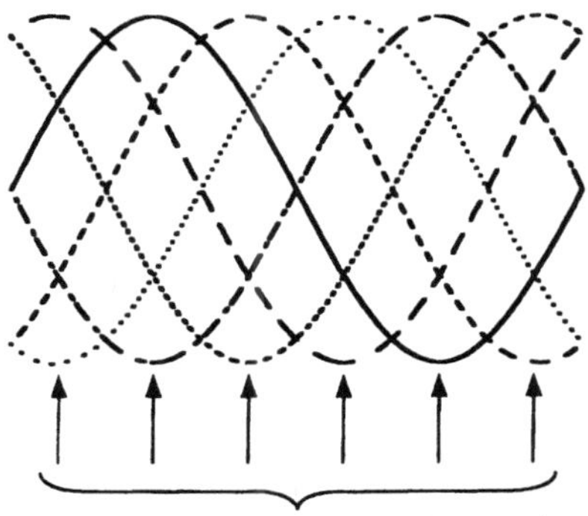

Figure 5.24. Waveforms showing the six equivalent states for a three stage oscillator

We will study the ISF for the power supply first. For an N-stage differential ring oscillator operating at frequency ω_0, it can be shown that its ISF is a periodic function with a frequency of $2N\omega$, i.e.,

$$\Gamma(t) = f(2N\omega_0 t). \tag{5.35}$$

Let us consider the three-stage differential ring oscillator as shown in Figure 5.23. Figure 5.24 illustrates the waveforms for the internal nodes A, A', B, B', C and C'. It is seen from Figure 5.24 that the oscillator goes through six equivalent states in a clock cycle, hence the ISF for V_{dd} has a frequency of $6\omega_0$. This can be generalized to an N-stage differential ring oscillator whose ISF for V_{dd} has a frequency of $2N\omega_0$.

Based on the analysis in [Hajimiri et al., 1999], only supply and substrate noise whose frequency is close to DC and $2iN\omega_0$ will make major contribution to the VCO phase noise where $i = 1, 2, 3, ...$ It can be shown that for supply noise close to DC, the SSB phase noise is given by Equation (5.36) where v_0 is the noise amplitude, Γ_0 is the DC term for Γ and $\Delta\omega$ is the noise frequency.

$$L(\Delta\omega) = \frac{v_0^2 \Gamma_0^2}{8\Delta\omega^2} \tag{5.36}$$

For supply noise close to $2iN\omega_0$, the SSB phase noise is given by Equation (5.37) where c_i is the ith Fourier coefficient for Γ and $\Delta\omega$ is the offset frequency from $2iN\omega_0$.

$$L(\Delta\omega) = \frac{v_0^2 c_i^2}{32\Delta\omega^2} \tag{5.37}$$

From Equations (5.36) and (5.37), it can be seen that, for supply noise either close to DC or $2iN\omega_0$, the resulting phase noise is inversely proportional to $\Delta\omega^2$. However, the previous statement is not precisely true when Γ_0 is a function of frequency. We shall provide more details of this discrepancy later in this section. In most high-frequency applications, the frequency $2iN\omega_0$ is at multiple GHz and above. However, the majority of the power supply noise is closer to DC than to any of the harmonic frequencies. As a result, the low-frequency supply noise has more of an impact on the ring oscillator phase noise, and Equation (5.36) can be used for most practical design considerations.

There are also applications where the clock jitter is more of a concern. It is beneficial to discuss the impact of the supply and substrate noise in terms of jitter. In particular, we will use the term *period jitter* δT which is the variation of the clock period.

With low-frequency supply noise $v_0 \sin(\Delta\omega t)$ where $\Delta\omega \ll 2N\omega_0$ the instantaneous frequency of oscillation $\omega(t)$ is given by

$$\omega(t) = \omega_0 + \omega_n(t) = \omega_0 + v_0\Gamma_0 \sin(\Delta\omega t) \tag{5.38}$$

Hence,

$$\frac{1}{\omega(t)} = \frac{1}{\omega_0 + v_0\Gamma_0 \sin(\Delta\omega t)} \approx \frac{1}{\omega_0}\left[1 - \frac{v_0\Gamma_0}{\omega_0}\sin(\Delta\omega t)\right] \tag{5.39}$$

As a result, the RMS value for the period jitter is defined by Equation (5.40). It is interesting to note that it is independent of the frequency of the supply and substrate noise $\Delta\omega$.

$$|\delta T|_{RMS} = \left|\frac{2\pi}{\omega} - \frac{2\pi}{\omega_0}\right|_{RMS} = \frac{v_0\Gamma_0}{\sqrt{2}\omega_0^2} \tag{5.40}$$

5.2 Simulation Results

Due to the presence of decoupling capacitors on the circuit board and the bond wire inductance, it is hard to precisely control and measure the ripple that the circuit sees on the power supply. Therefore, we use simulations results to verify our power supply noise model on a three-stage Maneatis ring oscillator. We introduce a sinusoidal ripple with a peak voltage of $0.1V$ added to the power supply for these simulations. Figure 5.25 shows the sideband power spectral density (PSD) as a function of the offset frequency Δf. In particular,

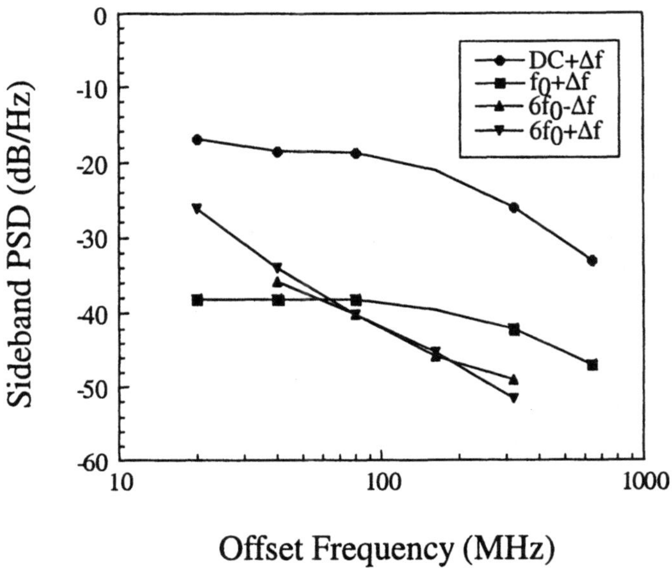

Figure 5.25. Sideband PSD when sinusoidal ripple is added to the power supply

the frequency of the ripple signal, f_r, is at Δf, $f_0 + \Delta f$ and $6f_0 \pm \Delta f$ where $f_0 = 903MHz$ and $\Delta f \ll f_0$.

It can be seen from the figure that, for frequencies close to DC and $6f_0$, the PSD for the phase noise drops at a rate of 20dB/dec as Δf increases for higher frequencies. The lower rate for frequencies less than 100 MHz is because the delay cell has a fully differential topology whose power supply rejection ratio (PSRR) improves at low frequencies. In other words, its Γ_0 becomes smaller at low frequencies. As a result, the curve for $DC + \Delta f$ stops increasing at low offset frequencies.

We have also simulated the oscillator with the ripple frequency, f_r, close to f_0. As shown in Figure 5.25, we do not see a $1/\Delta f^2$ slope as in the case of $DC + \Delta f$ and $6f_0 \pm \Delta f$. Instead, it only changes slightly over frequency. Additionally, we note that its absolute value is lower than for both $DC + \Delta f$ and $6f_0 \pm \Delta f$ for low offset frequencies.

In a practical CMOS PLL design, high frequency supply and substrate noise is usually more problematic due to reduced isolation at higher frequencies [Joardar, 1994]. In [Joardar, 1994] the noise power increases at a rate of 20dB/decade over frequency until about 500 MHz. If we neglect the impact of supply and substrate noise on the other components in a PLL and use the results in [Joardar, 1994] and Figure 5.25, the supply and substrate noise causes spurs beyond the PLL bandwidth until about 500 MHz with approximately equal amplitude. As a result, it is important to minimize Γ_0. Fully differential topologies improve the immunity to the supply and substrate noise at low frequencies. Unfortunately they do not help much at high frequencies as their PSRR degrades. The impact of supply and substrate noise on the other PLL components is beyond the scope of this book.

In general, N-stage differential ring oscillators are sensitive to power supply and substrate noise at frequencies close DC, $2N\omega_0$ and its harmonics. Most practical systems are bandlimited and therefore, the noise near DC is more important. Fully differential structures provide some common-mode rejection for lower frequencies. However, they have limited impact at higher frequencies. This increased rejection may have limited benefit because most VCOs are used within PLLs, where low frequency noise is suppressed within the loop bandwidth and high-frequency noise is more of a problem.

6. Conclusions

In this section, we have introduced a set of ring oscillator phase noise models which include the additive noise from the delay cells, up-converted noise from the bias and frequency control circuits and the supply/substrate noise. Our models are validated by simulations and measurement results. Our analysis suggests that fast rail-to-rail switching is needed to minimize ring oscillator phase noise. Current bias circuits have to be avoided if possible to reduce the up-converted low-frequency noise. The supply and substrate noise degrade the oscillator phase noise around DC and $2iN\omega_0$. Fully differential topologies improve PSRR at low frequencies, but they provide little help at high frequencies. We have developed phase noise models that not only predict ring oscillator phase noise, but also provide an intuitive direction for low-phase-noise oscillator design.

Notes

1 The $\Gamma(\omega_0 t)$ is a function that defines the impact on the output phase by a unit current impulse at time equal to t.

2 The f_T actually defines the frequency at which the current gain of the device is equal to one.

3 We use such large capacitor values because the filter corner frequency is set by g_{m6}/C_{BP} where C_{BP} is the by-pass capacitor value.

Chapter 6

NEW RING VCO DESIGN

1. Introduction

We have presented some theoretical work in the last chapter for ring oscillator phase noise analysis. In this chapter, we focus on practical design considerations. A ring VCO with differential control and quadrature outputs is taken as a example to demonstrate low phase noise VCO design. This chapter is organized as follows. First, we provide an overview of phase noise in ring oscillators in Section 2. Then we present our circuit design in Section 3. Then we extend our phase noise model to include delay cells with hysteresis in Section 4. In Section 5, we validate our design with some simulation and measurement results. Finally we draw some conclusions in Section 6.

2. Phase Noise Overview

In order to help understand our design considerations for low phase noise, we briefly review some theoretical analysis for ring oscillator phase noise in this section. Compared to Chapter 5, we will focus more on practical design considerations instead of theoretical analysis.

In a traditional ring oscillator design with fully differential delay cells, there are two major intrinsic causes for large phase noise [Dai and Harjani, 2000a], and they are illustrated in Figure 6.1.

One is the additive white noise from the differential pair and the loads. We view the phase noise as noise-to-carrier ratio, which is equivalent to the inverse of signal-to-noise ratio (SNR). In most designs with traditional fully differential delay stages, the signal level is limited to a few hundred millivolts, and therefore limits the achievable SNR. The single-sideband (SSB) phase noise due to the additive noise can be written in terms of the slew rate as in Equation (6.1) for a 3-stage ring oscillator [Dai and Harjani, 2000a].

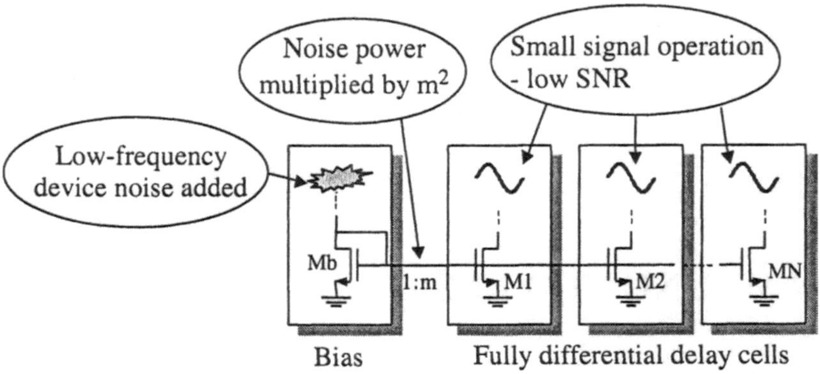

Figure 6.1. Illustration of major noise sources in a traditional fully differential ring oscillator

$$L(\Delta\omega) = \begin{cases} \frac{16FkTR\omega_0^2}{9|dv/dt|^2_{max}} \left(\frac{\omega_0}{\Delta\omega}\right)^2 & (for\ V_{pp} \ll \frac{8V_{dd}}{3\pi}) \\ \frac{64FkTRV_{dd}\omega_0^3}{27\pi|dv/dt|^3_{max}} \left(\frac{\omega_0}{\Delta\omega}\right)^2 & (for\ V_{pp} \gg \frac{8V_{dd}}{3\pi}) \end{cases} \tag{6.1}$$

The other major noise source in a traditional design is the up-converted low-frequency flicker noise from the bias circuit. The noise of each device in the bias circuit adds up and is mirrored into each delay stage. It modulates the frequency (FM) by changing the delay time instantaneously. Therefore, the low-frequency noise is up-converted to the vicinity of the carrier. To make matters worse, the current mirrors that supply the bias current usually have a ratio of m with $m > 1$ in order to save power. The bias noise is then amplified by m^2 in terms of power, hence the up-converted phase noise usually dominates the low-frequency noise. The phase noise due to the flicker noise is given by Equation (6.2) [Dai and Harjani, 2000a].

$$L(\Delta\omega) = \frac{\pi(\frac{1}{N} + m)K_f\omega_0^2}{8C_{ox}L_{eff}^2 I_{ss}^{2-A_f}\Delta\omega^3} \tag{6.2}$$

When integrated with the digital circuits, there is substantial switching noise from the digital circuits coupled through the power supply and substrate. In many cases, it causes spurs much higher than the intrinsic device noise. Therefore, it is important for PLLs to have good power supply and substrate noise rejection. Differential circuits are well known for their improved power supply rejection at low frequencies, and they are widely used for ring oscillator circuits. However, few of the published designs are controlled by differential voltages. In most PLL designs, either a single-ended charge pump and loop filter have to be used, or a differential-to-single-ended converter is inserted between the loop filter and the VCO. In either case the supply and substrate noise can couple to the control input and result in VCO phase noise. The noise in the differential-to-single-ended converter is equivalent to the bias noise. As we showed in the last chapter, its noise can be up-converted to phase noise. In particular, its flicker noise may dominate the low-frequency phase noise.

We have designed a two-stage CMOS ring oscillator with the emphasis of minimizing the phase noise due to the mechanisms discussed above. Its design and analysis will be presented in later sections in this chapter.

3. Circuit Design

In this section, we will present our ring VCO circuit diagram, and discuss our design considerations for the low phase noise and common-mode rejection.

Figure 6.2 shows the hierarchical schematics for the ring oscillator. The diagram in (a) is a top-level schematic. As shown in (b) and (c), each delay cell has complementary inputs, complementary outputs and differential control terminals. We have introduced a latch in the delay cell consisting of $X3$ and $X4$, which can store either a logic 1 or a 0.

The input signals drive a pair of inverters, $X1$ and $X2$, whose strength is controlled by the differential voltage of V_c^+ and V_c^-. If the input flips over, after a certain delay, the output will flip over, too. The delay time is determined by the relative strength ratio between the input inverters $X1/X2$ and the latch $X3/X4$. The stronger $X1$ and $X2$ are, the less delay time is needed for the output to change its state. In Figure 6.2, NMOS $M1$ and PMOS $M2$ are introduced to adjust the strength of $X1$ and $X2$, hence to change the delay time.

In most applications, the frequency tuning range for a ring oscillator is more than sufficient. Therefore, its gain should be minimized in order to reduce the phase noise [Wilson et al., 2000]. Figure 6.3 illustrates the channel resistance of NMOS devices with $V_{ds} = 0$ as a function of the gate voltage. For the same channel resistance value, a narrower device needs a larger gate voltage. The gate voltage has to be larger to bias the device more into the triode region. Also a larger V_{gs} reduces the control sensitivity and improves the linearity. Therefore, the voltage for V_c^+ should be close to V_{dd} and the voltage for V_c^- should be close to ground, so that both $M1$ and $M2$ are biased in deep triode.

Since $M1$ and $M2$ have the opposite control polarities, they can be sized properly such that the common-mode voltage at V_c^+ and V_c^- is cancelled out and only the differential control signal determines the frequency of oscillation. Since a large portion of the interference coupled from the power supply and the substrate is common mode, our design has good supply and substrate noise rejection/indexsupply/substrate noise rejection. This is an important feature when the VCO is integrated on the same silicon substrate with many other digital circuits.

A simply way to achieve the common-mode cancellation is to size $M1$ and $M2$ so that they have the same strength which is expressed in Equation (6.3).

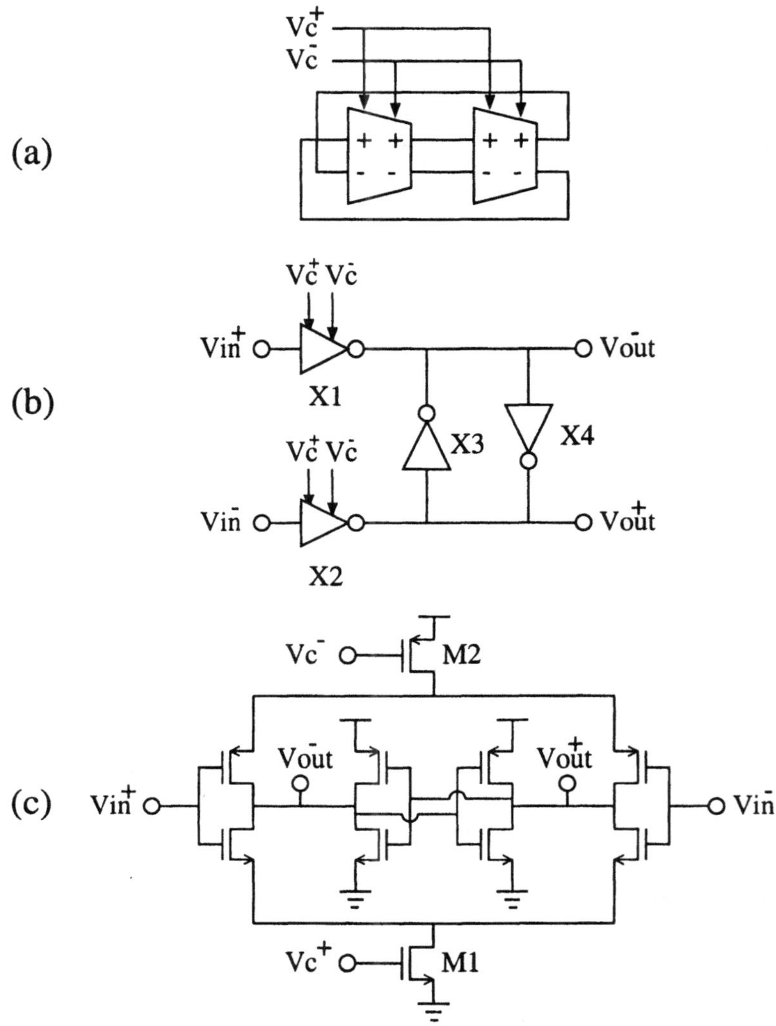

Figure 6.2. VCO schematic (a) top-level (b) logic-level for delay cells (c) transistor-level for delay cells

Also the common-mode control voltage should be set at the voltage give by Equation (6.4). Here, μ is the mobility of the devices which includes the effects of mobility degradation for short channel devices. W and L are the width and length of the devices respectively. V_T is the transistor threshold. The subscripts N and P are used to represent NMOS and PMOS transistors.

$$\frac{\mu_N W_N}{L_N} = \frac{\mu_P W_P}{L_P} \tag{6.3}$$

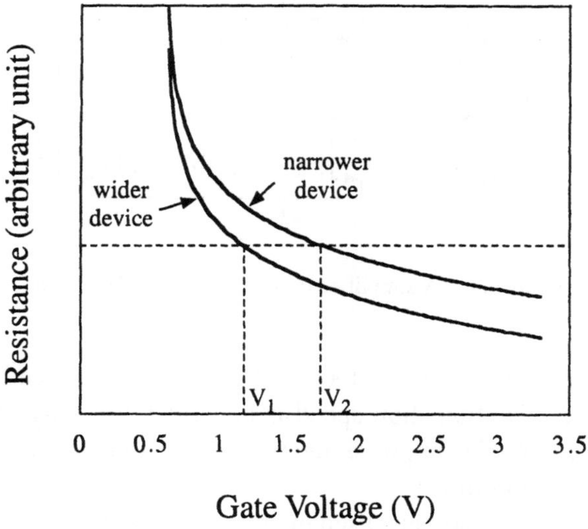

Figure 6.3. Channel resistance of an NMOS transistor with $V_{ds} = 0$ vs. gate voltage.

$$V_{cm} = \frac{V_c^+ + V_c^-}{2} = \frac{V_{dd} + V_{TN} - |V_{TP}|}{2} \tag{6.4}$$

A typical ring oscillator needs at least three stages to oscillate. In our design, because there is an extra delay introduced in each stage, a two-stage design is sufficient to ensure oscillation [Razavi, 1997]. Since each stage contributes phase noise, our design has fewer noise sources than a regular design with three or more stages.

It can be shown that the phase noise for our circuit is given by Equation (6.5), where $L(\Delta\omega)$ is single-sideband phase noise, α is a empirical constant factor, F is excess noise factor introduced in Leeson's model [Leeson, 1966], k is the Boltzmann's constant, T is the absolute temperature, ω_0 is the center frequency, $\Delta\omega$ is the offset frequency, V_{pp} is the internal peak-to-peak signal voltage and V_{dd} is the power supply voltage [Dai and Harjani, 2000a].

$$L(\Delta\omega) = \frac{\alpha F k T R V_{dd} \omega_0^3}{|dv/dt|_{max}^3} \left(\frac{\omega_0}{\Delta\omega}\right)^2 \tag{6.5}$$

In our design, when the latch changes its state, it forms a positive feedback system which speeds up the transition. Inverters $X3/X4$ are made slightly stronger than $X1/X2$. The transition time is primarily determined by $X3/X4$ and it is relatively independent of the frequency of oscillation. It is important to keep the frequency tuning devices $M1$ and $M2$ outside the positive feedback loop so that the transition speed is maximized for a given technology. The

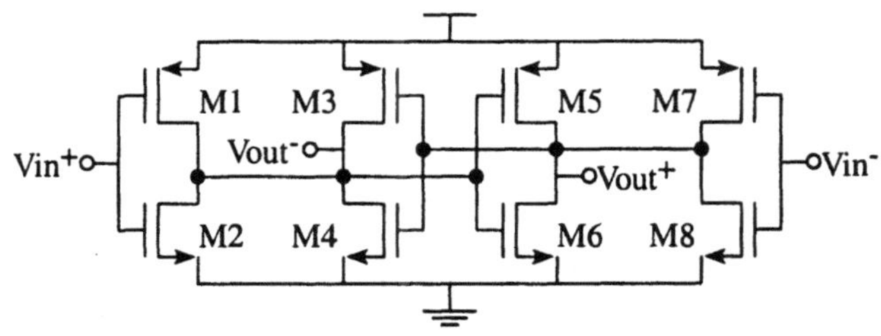

Figure 6.4. A simplified delay cell without delay control

oscillator is tuned by changing the delay without significantly affecting the transition speed. The transition speed is governed primarily by the f_T of the process. As technologies improve, the transition speed will increase, reducing the phase noise even further.

Even though a latch has been used in previous ring oscillator designs [Thamsirianunt and Kwasniewski, 1997], the purpose of the latch was simply to synchronize the two inverter rings in order to generate the complementary outputs. The idea in our design is to maximize the transition speed and is therefore clearly distinguishable from the previous work.

4. Analysis of Circuits with Hysteresis

A major difference between our circuit and traditional ring oscillator designs is that our circuit includes delay cells with hysteresis. In other words, the output of the delay stage is not only a function of its input voltage, but is also dependent on its history. Our analysis in this section will include hysteresis. This will help us understand our circuit better, and it can also be applied to future designs with hysteresis delay cells.

We have simplified the delay cells in our design as shown in Figure 6.4. Compared to Figure 6.2, the transistors $M1$ and $M2$ have been removed, since their existence is equivalent to changing the widths of $M1$, $M2$, $M7$ and $M8$ in Figure 6.4.

Now let us look at the step response of the delay stage. Figure 6.5 shows an example case when the initial voltages for V_{out}^+ and V_{out}^- are V_{dd} and 0 respectively and V_{in}^+ and V_{in}^- take a negative and positive step respectively. All the off-devices have been removed to simplify the circuit schematic. As a response to the step at the inputs, currents start to flow as indicated by the arrows in the figure, trying to flip over the output voltages.

Now we will compute the input threshold voltage V_{th} that causes the outputs to flip over. As V_{in}^+ falls and V_{in}^- rises, the output remains at its initial state until $M3$ or $M6$ turns on. At that point, the positive feedback formed by the cross coupled NMOS and PMOS pairs becomes active, and the output changes its state. Assuming the threshold point is indicated by $V_{in}^+ = V_{dd} - V_{th}$ and

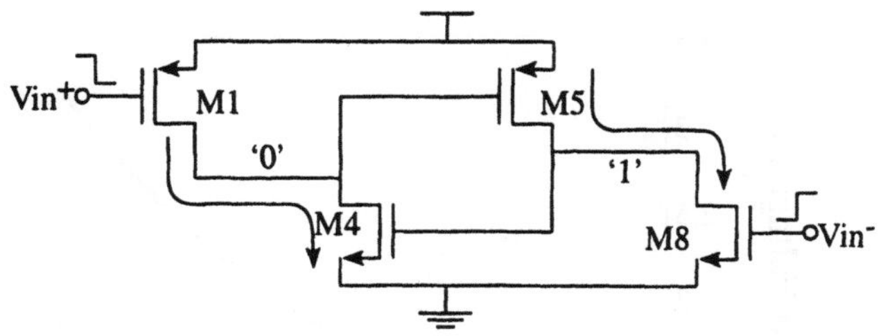

Figure 6.5. An example case for a step response

$V_{in}^{-} = V_{th}$, the output voltages are $V_{out}^{+} = V_{dd} - |V_{TP}|$ and $V_{out}^{-} = V_{TN}$. Normally the NMOS and PMOS transistors are not turned on at the same time, because NMOS transistors have lower threshold voltages then PMOS and they are turned on earlier. However, because of the positive feedback, once the NMOS positive feedback becomes active, it will change the voltages drastically and force the PMOS transistors into its own positive feedback. Therefore, we can assume that both V_{out}^{+} and V_{out}^{-} meet their thresholds together. For $3.3V$ power supply, the above conditions mean that at the threshold, $M1$ and $M8$ are in saturation and $M4$ and $M5$ are in triode.

For simplicity, let us assume that $M1/M7$ have equal strength with $M2/M8$, and $M3/M5$ have equal strength with $M4/M6$, i.e.,

$$\beta_1 = \frac{\mu_N W_1}{L_1} = \frac{\mu_N W_7}{L_7} = \frac{\mu_P W_2}{L_2} = \frac{\mu_P W_8}{L_8} \tag{6.6}$$

$$\beta_2 = \frac{\mu_N W_3}{L_3} = \frac{\mu_N W_5}{L_5} = \frac{\mu_P W_4}{L_4} = \frac{\mu_P W_6}{L_6} \tag{6.7}$$

The KCL equation at node V_{out}^{+} can be written as

$$\frac{\beta_1}{2}\left(V_{th} - V_{TN}\right)^2 = \beta_2\left[\left(Vdd - V_{TN} - |V_{TP}|\right)|V_{TP}| - \frac{1}{2}V_{TP}^2\right] \tag{6.8}$$

Solving Equation 6.8, the threshold voltage is given by

$$V_{th} = V_{TN} + \sqrt{\frac{2\beta_2}{\beta_1}\left[\left(V_{dd} - V_{TN} - |V_{TP}|\right)|V_{TP}| - \frac{1}{2}V_{TP}^2\right]} \tag{6.9}$$

A plot for V_{th} vs. β_2/β_1 is shown in Figure 6.6. Interestingly the output can change states even when $\beta_2/\beta_1 > 1$, which means that the input inverters are weaker than the latch. This is because of the fact that the input driving transistors are in the saturation region while the transistors in the latch are in either off or triode region. Therefore, the driving capability of the input devices is at its maximum while the devices in the latch are more like resistors that have less control of its own state.

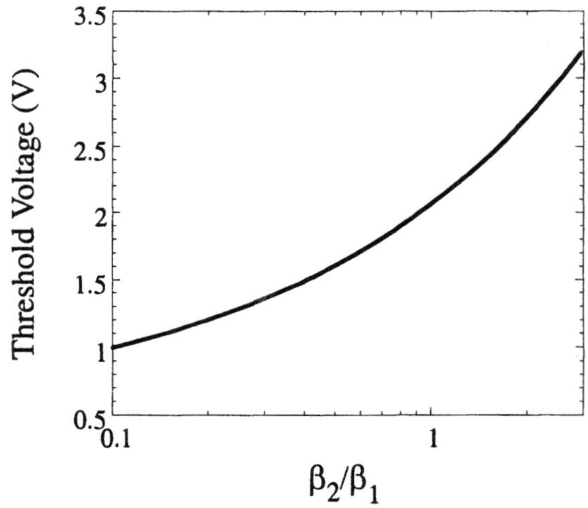

Figure 6.6. The threshold of the delay cell as a function of β_2/β_1

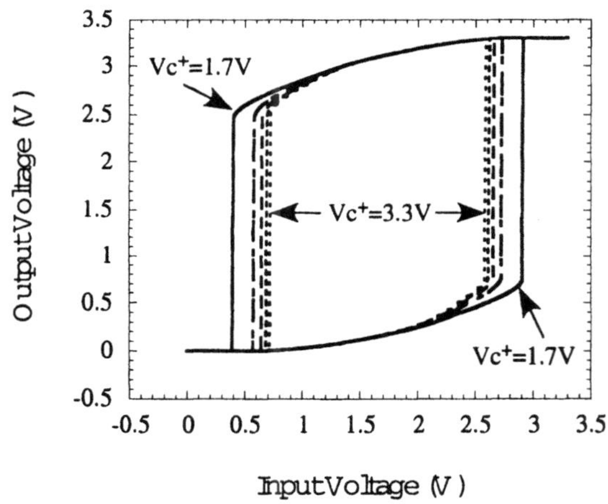

Figure 6.7. Simulated hysteresis curve for the delay cell

The hysteresis curve of the delay cell is simulated and is shown in Figure 6.7. Changing the control voltages V_c^+ and V_c^- is equivalent to changing the strength of devices in the input inverters. In Figure 6.7, only V_c^+ is labelled. V_c^- can be implied from the fixed common-mode voltage of $1.25V$. Similarly only the positive sides of input and output voltages are plotted. It can be seen from the figure that as the differential control voltage decreases, the hysteresis loop becomes wider. This is consistent with Equation 6.9.

Now we are ready to study a ring oscillator with hysteresis delay cells. An N-stage ring oscillator with such delay cells is drawn in Figure 6.8. Resistance

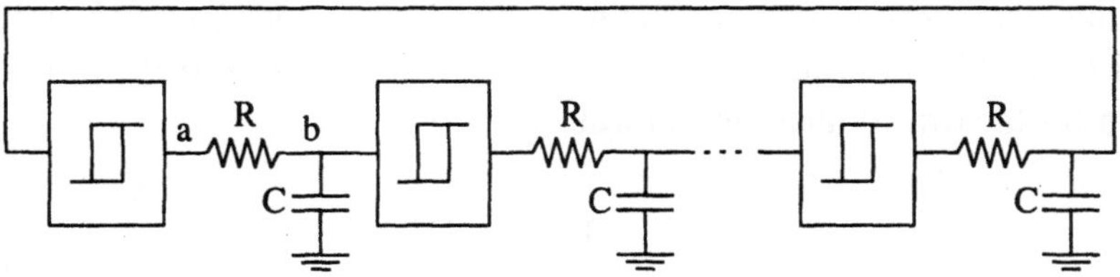

Figure 6.8. An N-stage ring oscillator with hysteresis delay cells

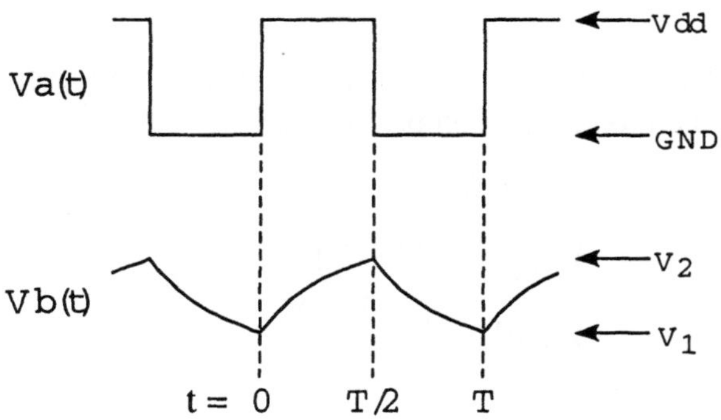

Figure 6.9. Waveforms at nodes a and b in Figure 6.8

R and capacitance C are added to each stage to model the output impedance and the input parasitic capacitance of the delay cell.

The waveforms at the nodes a and b are illustrated in Figure 6.9. The hysteresis delay cell generates a square wave at node a, which swings between ground and V_{dd}. The resistor R and capacitor C provide a first-order low-pass shaping of the waveform at node b.

Let us assume that the initial voltage for each rising edge at node b is V_1 and the initial voltage for each falling edge is V_2 as shown in Figure 6.9. Then $V_b(t)$ is given by Equation 6.10.

$$V_b(t) = \begin{cases} V_{dd}\left(1 - e^{-\frac{t}{RC}}\right) + V_1 e^{-\frac{t}{RC}} & (for \quad 0 < t < \frac{T}{2}) \\ V_2 e^{-\frac{t-\frac{T}{2}}{RC}} & (for \quad t > \frac{T}{2}) \end{cases} \tag{6.10}$$

With the periodic boundary conditions $V_a(T/2) = V_2$ and $V_b(T) = V_1$, Equation 6.10 can be rewritten as in Equations 6.11 and 6.12.

$$V_{dd}\left(1 - e^{-\frac{T}{2RC}}\right) + V_1 e^{-\frac{T}{2RC}} = V_2 \tag{6.11}$$

$$V_2 e^{-\frac{T}{2RC}} = V_1 \tag{6.12}$$

Solving above equations, we obtain the expressions for V_1 and V_2 in Equations 6.13 and 6.14., which, when combined with Equation 6.10, provides a solution for the time domain waveform.

$$V_1 = \frac{V_{dd}}{1 + e^{\frac{T}{2RC}}} \tag{6.13}$$

$$V_2 = \frac{V_{dd} e^{\frac{T}{2RC}}}{1 + e^{\frac{T}{2RC}}} \tag{6.14}$$

At $t = T/(2N)$ after V_b starts rising from V_1, it must cross over V_{th} to trigger the next stage. Therefore, Equation 6.15 can be derived.

$$\begin{aligned} V_{th} &= V_b\left(\frac{T}{2N}\right) \\ &= V_{dd}\left(1 - e^{-\frac{T}{2NRC}}\right) + \frac{V_{dd}}{1+e^{\frac{T}{2RC}}} e^{-\frac{T}{2NRC}} \\ &= V_{dd}\left(1 - \frac{e^{-\frac{T}{2NRC}}}{1+e^{-\frac{T}{2RC}}}\right) \end{aligned} \tag{6.15}$$

For a two-stage oscillator, $N = 2$, Equation 6.15 gives

$$T = 4RC \ln\left[\frac{V_{dd} + \sqrt{(3V_{dd} - 2V_{th})(2V_{th} - V_{dd})}}{2(V_{dd} - V_{th})}\right] \quad \left(for\ \frac{1}{2}V_{dd} < V_{th} < V_{dd}\right) \tag{6.16}$$

The condition $\frac{1}{2}V_{dd} < V_{th} < V_{dd}$ means that the delay cell has hysteresis, without which the two-stage circuit would not oscillate. As V_{th} approaches $\frac{1}{2}V_{dd}$, the oscillation period T approaches 0. This means that as the hysteresis diminishes, the oscillator runs faster and faster until it reaches the frequency at which the active device cannot provide any gain and the oscillation has to stop. As V_{th} approaches V_{dd}, the delay cells become harder and harder to change the states, causing the period to approach infinity. A plot for $T/(RC)$ vs. V_{th}/V_{dd} is shown in Figure 6.10. It is seen that the period of the oscillation approaches 0 as V_{th} approaches $0.5V_{dd}$, i.e., the hysteresis disappears. This implies that practically the oscillator will stop oscillation since the devices can no longer provide enough gain at higher and higher frequencies as the hysteresis diminishes. This is consistent with the well-known conclusion that a two-stage ring cannot sustain oscillation. Typically each stage is a single-pole low-pass system, which provide no more than 90° of phase shift. Therefore, at least three stages are needed to obtain 180° of phase shift at a finite frequency. A two-stage become possible only when the delay cells have hysteresis which adds additional delay to meet Barkhausen's criteria.

The circuit noise can be viewed as equivalent to a variation in the threshold V_{th}. Therefore, in order to minimize phase noise, the value of dT/dV_{th} has to

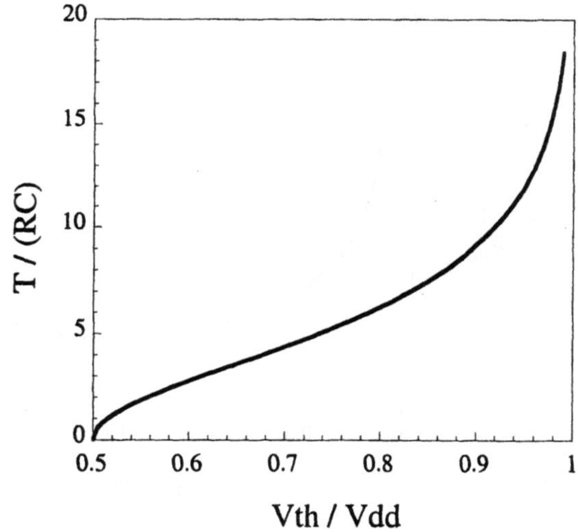

Figure 6.10. $T/(RC)$ as a function of V_{th}/V_{dd} for a 2-stage ring oscillator

be minimized. To simplify the math, let $\gamma = V_{th}/V_{dd}$. Then for fixed R, C and V_{dd}, the sensitivity of T with respect to V_{th} is given by Equation 6.17.

$$\frac{dT}{dV_{th}} = \frac{4RC}{V_{dd}}\left[\frac{8(1-\gamma)}{(3-2\gamma)(2\gamma-1)+\sqrt{(3-2\gamma)(2\gamma-1)}} + \frac{1}{1-\gamma}\right] \tag{6.17}$$

However, a more important figure of merit is $\frac{dT}{dV_{th}}/\frac{T}{V_{dd}}$, since it represents the amount of jitter relative to the period. Furthermore, it is a dimensionless quantity that is only a function of γ. This can be obtained by combining Equations 6.16 and 6.17 for two-stage oscillators and its expression is given by Equation 6.18.

$$\frac{dT}{dV_{th}}/\frac{T}{V_{dd}} = \frac{\frac{8(1-\gamma)}{(3-2\gamma)(2\gamma-1)+\sqrt{(3-2\gamma)(2\gamma-1)}} + \frac{1}{1-\gamma}}{\ln\left[\frac{1+\sqrt{(3-2\gamma)(2\gamma-1)}}{2(1-\gamma)}\right]} \tag{6.18}$$

Its values are plotted in Figure 6.11. It is seen from the plot that the minimal sensitivity is achieved for $\gamma \approx 0.81$ and the minimum value is 3.77. However, the curve in the plot remains relatively flat for $0.7 < \gamma < 0.9$, which means that the phase noise does not increase dramatically when γ is in this region. As γ approaches 0.5 or 1, the sensitivity becomes infinitely large. For a $3.3V$ power supply, this corresponds to a threshold region between $2.31V$ and $2.97V$, with $2.66V$ being the optimal point. From Figure 6.7 it is seen that V_{th} for our design

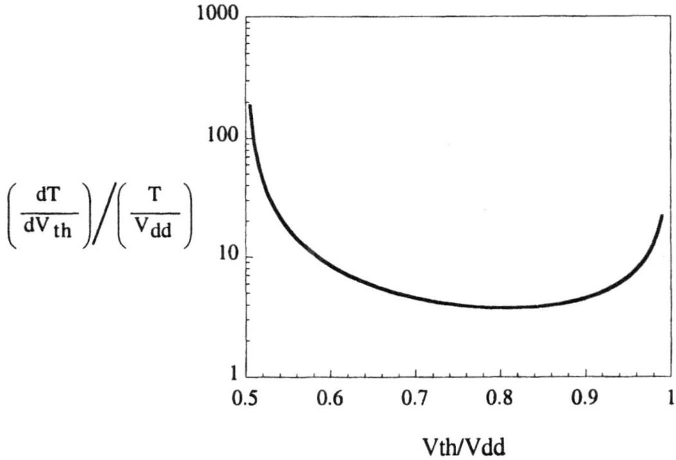

Figure 6.11. Sensitivity of period with respect to threshold voltage for a 2-stage ring oscillator

falls in this optimal range. As V_{th}/V_{dd} approaches 0.5, the curve in Figure 6.11 loses its practical meaning, since the circuit will stop oscillation. On the other hand, as V_{th}/V_{dd} approaches 1, the curve goes to infinity. This means that the circuit will become very jittery, since it is increasingly sensitive to circuit noise if the delay cells have too much hysteresis.

We have also done similar analysis for a 3-stage ring oscillators with hysteresis. Its mathematical expression is too complicated to give any insight into the understanding of the phase noise analysis. Therefore, the close-form expression is not derived in this book. Figure 6.12 shows $T/(RC)$ as a function of V_{th}/V_{dd} for the 3-stage ring oscillator. A major difference between this plot and the plot in Figure 6.10 for a 2-stage oscillator is that the period for a 3-stage oscillator does not go to 0 for $V_{th}/V_{dd} = 0.5$. This is intuitive and consistent with the fact that a 3-stage circuit can oscillate without any hysteresis.

The quantity of $\frac{dT}{dV_{th}}/\frac{T}{V_{th}}$ for a 3-stage ring oscillator is plotted in Figure 6.13. Its minimum is found at $V_{th}/V_{dd} = 0.73$ and the minimum value is 3.06. Compared to its value of 9.73 at $V_{th}/V_{dd} = 0.5$ as the case without hysteresis, the noise sensitivity can drop by 10dB when the optimal amount of hysteresis is added to the delay cells.

Comparing a three-stage oscillator and a two-stage oscillator, we see that less hysteresis is needed for a three-stage oscillator. Actually it can be induced that the more number of stages, the less hysteresis is needed to minimize phase noise. This can be understood as follows.

Figure 6.14 shows some example waveforms for ring oscillators with different number of delay stages operating at the same frequency. As the number of stages increases, the transitions have to take less time to maintain the same fre-

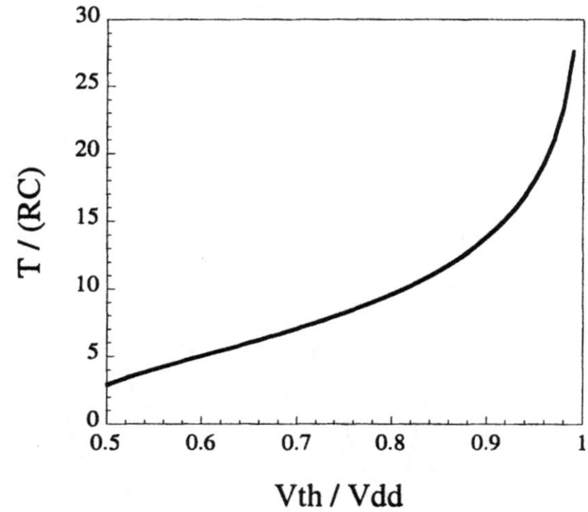

Figure 6.12. $T/(RC)$ as a function of V_{th}/V_{dd} for a 3-stage ring oscillator

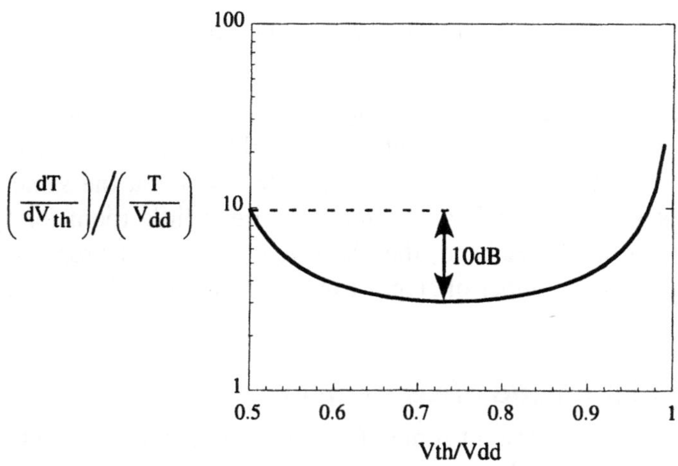

Figure 6.13. Sensitivity of period with respect to threshold voltage for a 3-stage ring oscillator

quency of oscillation. This corresponds to higher Q_{eff} as we have discussed in the previous chapter. However, more stages mean more noise sources. Therefore, it does not help much by using more stages for ring oscillators. Adding hysteresis to the delay cells effectively sharpens the transitions and makes the voltages stuck at V_{dd} and ground for a longer time during each period. It makes the waveform for a three-stage oscillator more like that for a four stage oscillator without adding another stage. Therefore, it improves Q_{eff} without increasing the number of noise sources. Hence, the oscillator phase noise can be reduced.

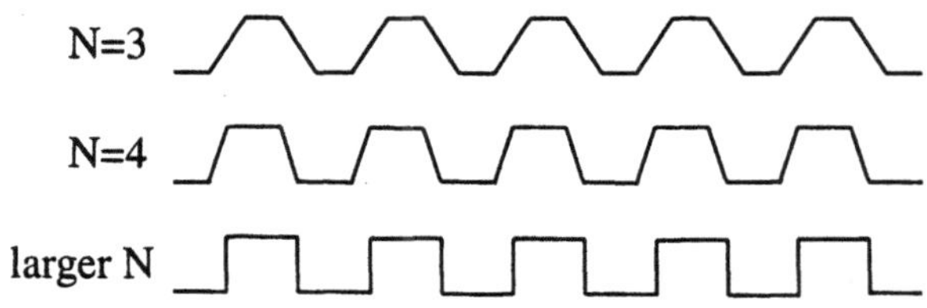

Figure 6.14. Waveforms for ring oscillators with different number of stages

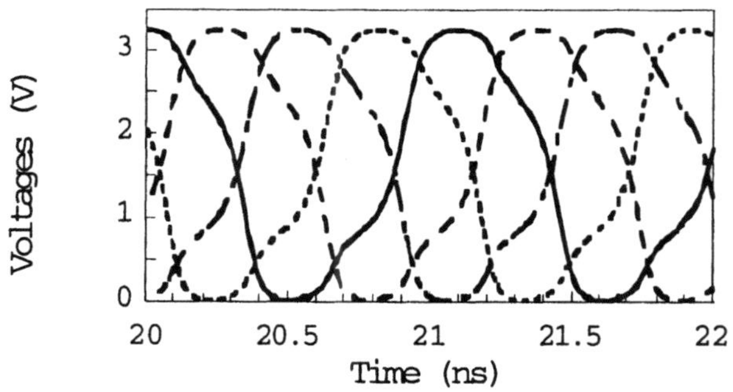

Figure 6.15. Quadrature output waveforms

This method is less effective for ring oscillators with large number of stages. As shown in Figure 6.14, the waveforms resemble square waves for large N's. The transition edges have to be very sharp to operate at the same frequency as oscillator with smaller N's. Therefore, there is little room for Q_{eff} to be improved. On the other hand, the phase noise is almost proportional to N. Thus, reducing the number of stages can result in lower phase noise for large N's.

5. Simulation and Measurement

In this section, we present some of our simulation and measurement results for our CMOS ring oscillator in order to validate our discussions in the previous sections.

Figure 6.15 shows the simulated quadrature outputs. Measured quadrature outputs are not provided due to loading constraints. Once the decision to flip over has been made, the positive feedback becomes active, and a fast transition occurs.

In a traditional ring oscillator design, any even number of stages can generate quadrature outputs, and four is the minimum qualified number. However, fewer number of stages are needed to achieve faster oscillation speed for a given process. In our design, the quadrature outputs are generated with a two-stage ring oscillator, which can potentially operate at a much higher frequency than a traditional design. As a result, it is more suitable for high-speed applications.

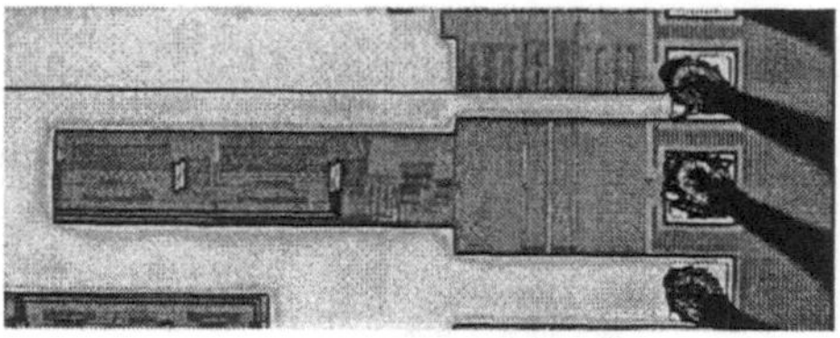

Figure 6.16. Die photo

Figure 6.17. Circuit board for VCO testing

The design has been fabricated in a TSMC $0.35\mu m$ CMOS process. The chip die photo is shown in Figure 6.16. It works with a 3.3V power supply and consumes 24mA current at 900MHz. The test circuit board is shown in Figure 6.17. All the components are surface-mounted to ensure high-frequency performance.

Figure 6.18 shows the measured DC control characteristic of the VCO. The top part of the curve is the desired region of operation. The common mode and the differential mode in the control voltages cause movement along the X-axis and the Y-axis respectively. Since the partial derivative of the curve along X-axis is close to zero, changes in the common mode control voltages do not change its frequency. The frequency is only controlled by the differential voltage. For common-mode AC signals at the control terminals, it will cause some instantaneous phase shift even though it does not change the actual frequency. Common mode noise is thus not removed completely, but is suppressed substantially.

To validate the impact of differential vs. single-ended control, we have measured the spectrum of the VCO output with a 75mV 700kHz interference at the control terminals. The test setup is shown in Figure 6.19. The test results are shown in Figure 6.20. The solid curve is the spectrum when the interference exists at both V_c^+ and V_c^-. The dashed curve is the spectrum when the interference exists only at V_c^+. The sideband is suppressed by 32dB at

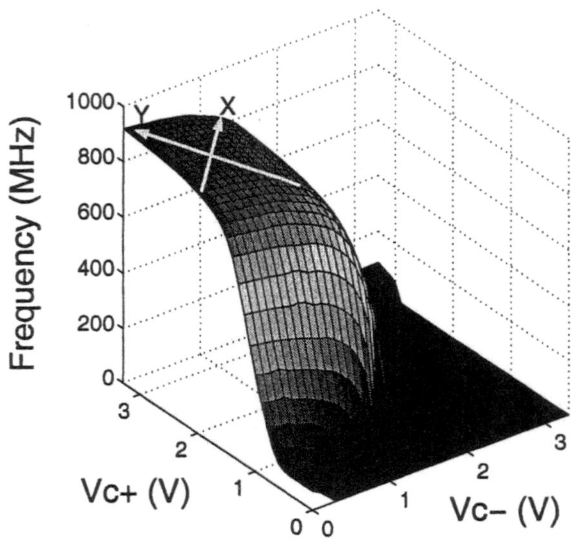

Figure 6.18. Measured control characteristic for the ring VCO

700kHz with the differential control topology. The interference of the higher harmonics are suppressed even further and cannot be observed in the plot. Figure 6.21 shows the sideband amplitude as a function of the frequency of the interference. When the interference appears only as common mode, there is no frequency modulation (FM), and only phase modulation (PM) exists. It can be shown that PM causes flat sideband as shown by the curve with solid circles and FM causes a sideband with 20dB/dec slope as shown by the curve with solid squares. Since the mismatch of the control devices results in a non-zero derivative along X-axis in Figure 6.18, the weak FM causes the sideband to increase slightly at low frequencies even with common-mode injection.

When the VCO is integrated with other digital circuits, there could be substantial digital switching noise coupled through the shared power supply and substrate. Therefore, it is important for the VCO to have good supply and substrate noise rejection. A conventional solution to this problem is to use fully differential delay stages with differential pairs. Ideally the source of the devices provides a virtual ground with the presence of supply and substrate voltage variation. This is shown in Figure 6.22 (a). Unfortunately this is no longer true for high-frequency noise. As shown in Figure 6.22 (b), the parasitic capacitance C_p due to the source/drain capacitances shorts the common source to ground at high frequencies, hence degrade the supply and substrate noise rejection.

In our design we do not have the differential pairs in our delay cells. The variation in the supply voltage may change the output conductance of transistors $M1$ and $M2$. Fortunately in our design the $|V_{gs}|$'s for $M1$ and $M2$ are large

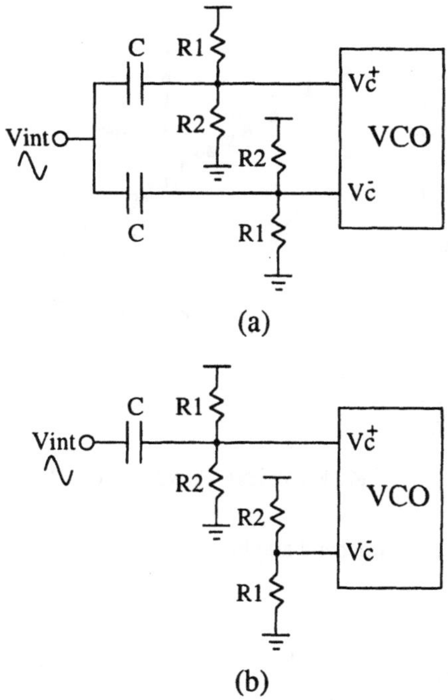

Figure 6.19. Test setup for common mode rejection at the control terminals (a) common-mode interference (b) single-ended interference

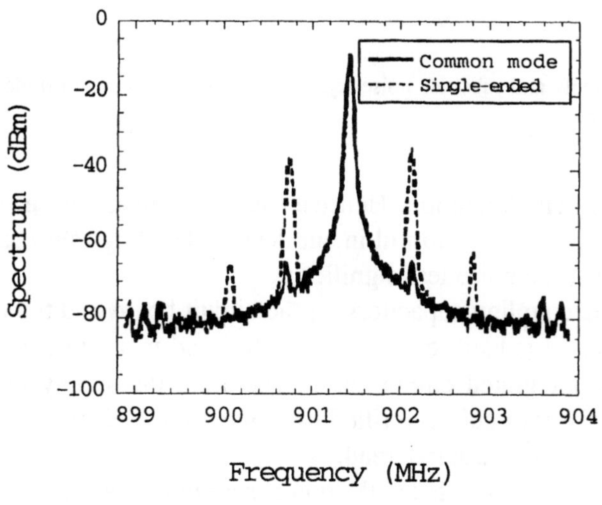

Figure 6.20. Spectrum with interference at the control terminals

since both of them are in deep triode region. Therefore the same amount of power supply ripple that exists on a conventional differential delay cell now only

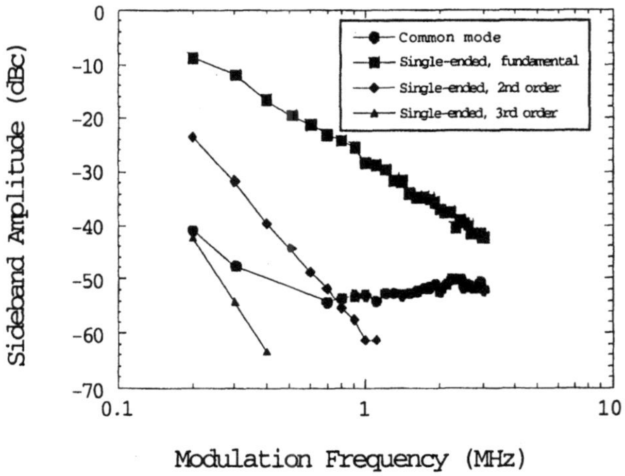

Figure 6.21. Sideband amplitude vs. frequency of interference

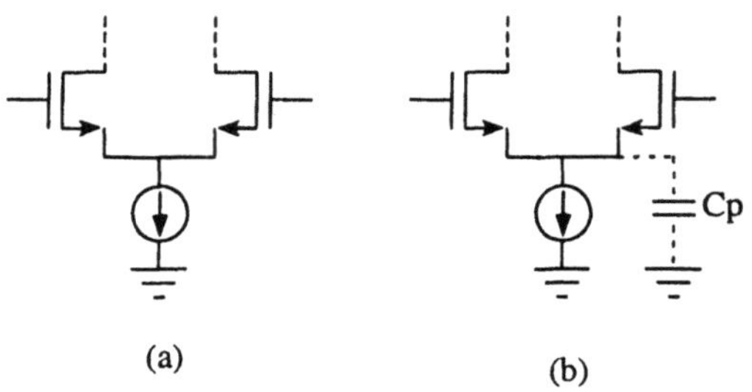

Figure 6.22. Supply and substrate noise rejection by differential pairs (a) ideal case (b) with parasitic capacitance

changes V_{gs} by a small fraction. Hence it causes less phase noise. Furthermore, the signal amplitude is rail-to-rail in our design. The large slew rate makes the supply and substrate noise less significant.

Due to the decoupling capacitors on the circuit board and the on-chip bond wire inductance, it is hard to precisely control and measure the power supply ripple that the integrated circuit sees. Therefore the supply noise rejection performance of our circuit is verified by simulation. Substrate noise rejection is expected to follow the same trend.

A sinusoidal ripple is added to the power supply with a peak voltage of 0.1V. As we have discussed in the previous chapter, the VCO is only sensitive to supply noise at $2iNf_0 (i = 0, 1, 2, \ldots)$ where N is the number of delay stages and f_0 is the center frequency of oscillation. The solid curves in Figure 6.23 show our simulation results for the sideband amplitude vs. the offset frequency from DC and $4f_0$ respectively. As a comparison, the simulation results for a

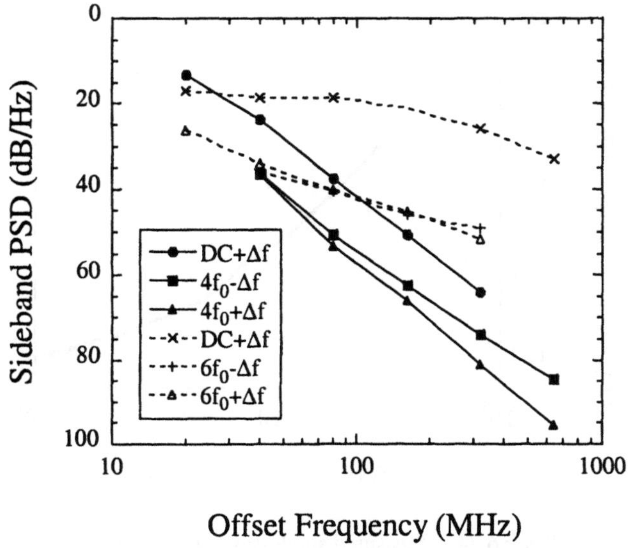

Figure 6.23. Sideband amplitude vs. frequency for supply noise

3-stage fully differential ring oscillator with Maneatis loads [Maneatis, 1996] is included in Figure 6.23 as dashed curves. Due to the arguments mentioned earlier, the frequency of the supply ripple is chosen to be around DC and $6f_0$.

For the Maneatis ring oscillator, as Δf decreases, the sideband rises at a rate of 20dB/dec until it flattens at low frequencies where the traditional differential pairs provide good power supply noise rejection. In our design, the sideband has a slope of 40dB/dec instead. Even though it does not provide power supply rejection near DC, its sideband is much lower than that for a traditional design at frequencies greater than 30MHz. At a frequency of 320MHz, an improvement of 38dB can be observed according to our simulation results in Figure 6.23.

For a digital circuit with random switching activities, the spectrum of the substrate noise is broadband and can extend from MHz to GHz range. The exact spectrum depends on the digital clock frequency, and circuit transition speed and the circuit topology [Xu et al., 2001]. However, the power of the substrate noise can be effectively reduced at low frequencies by substrate guard rings [Joardar, 1994]. Therefore, even though the exact shape of the substrate noise spectrum is unpredictable in a real digital environment, it is likely to focus most of its energy in the multiple hundred MHz to above 1GHz range. Our design has superior immunity to supply and substrate noise over traditional designs in this frequency range. Very low-frequency substrate noise should be partially suppressed by the PLL feedback loop.

The single-sideband phase noise was measured with a HP E4007B spectrum analyzer and LabVIEW. The result is shown in Figure 6.24. The oscillator was

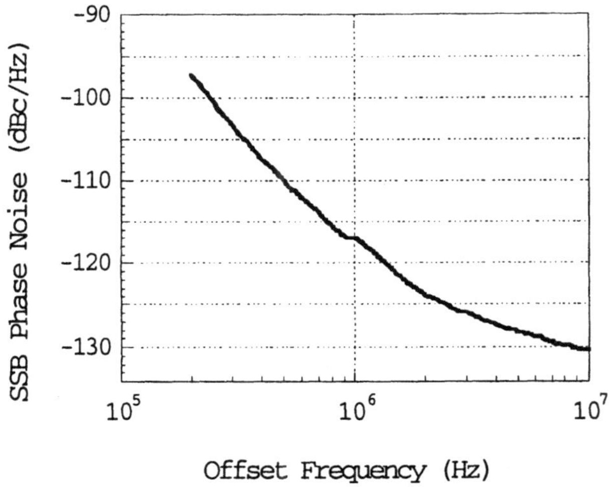

Figure 6.24. Measured single-sideband phase noise

operated at 973MHz for phase noise measurement. The measured phase noise is -117dBc/Hz at a 1MHz offset frequency. This is comparable to the measured phase noise for LC oscillators [Craninckx and Steyaert, 1995, Craninckx and Steyaert, 1997, Kinget, 1998].

6. Conclusions

We have designed a two-stage ring oscillator in a CMOS technology. In order to minimize phase noise, positive feedback is introduced in each delay cell to achieve fast switching speed. Differential frequency tuning is implemented in order to reject the common-mode interference on the control inputs. Its measured phase noise is -117dBc/Hz at a 1MHz offset frequency from a 973MHz carrier. This is comparable to typical LC-tank oscillators. The measured common-mode rejection shows 32dB improvement over traditional single-ended control topologies. The simulation results also indicate that our design has superior supply and substrate noise rejection over traditional designs. Our new VCO topology has low phase noise from intrinsic devices, and shows good common-mode rejection of interference from the control inputs, from the power supply and from the substrate.

We have also provided phase noise analysis for ring oscillators with hysteresis delay cells. It is shown that the phase noise can drop by 10dB when the optimal amount of hysteresis is added to the delay cells for a 3-stage ring oscillator.

Chapter 7

PLL DESIGN EXAMPLES

We have developed phase noise models for ring oscillators and applied our theoretical analysis to an example design of a ring oscillator. In a real system, a VCO is rarely used as a stand-alone component, because its output frequency is not well controlled. It is usually used in a PLL so that its output frequency is digitally controlled and its close-in phase noise is significantly suppressed by the PLL feedback loop.

As an example, we present two PLL designs in this chapter. They both target an output clock at 2.4GHz with an input reference clock at 1 MHz. Both PLLs use a fully differential charge pump and loop filter. One of the them has a ring VCO while the other one has an LC VCO. Their power consumptions are kept approximately equal.

As we have discussed in Chapter 2, the loop bandwidth K_0 cannot be greater than 1/10 of the PFD update rate. Otherwise, it will result in the overshoot in the frequency response or even cause the loop to become unstable. On the other hand, the loop bandwidth should be maximized in order to suppress the high-frequency phase noise from the VCO and reduce the lock-in time. Now if the PFD is updated at the same rate as the input clock which is at 1 MHz, the optimal loop bandwidth should be 100kHz.

Both PLLs were designed to meet Bluetooth specifications. Bluetooth protocol allows $220\mu s$ for the frequency synthesizer to settle each time when it switches to a new channel. The pull-in time for a PLL with a PFD is approximated by Equation 7.1 where f_{eo} is the initial frequency error, N is the feedback divider ratio and K is the loop bandwidth [Wolaver, 1991].

$$T_p \approx \frac{8f_{eo}}{NK^2} \tag{7.1}$$

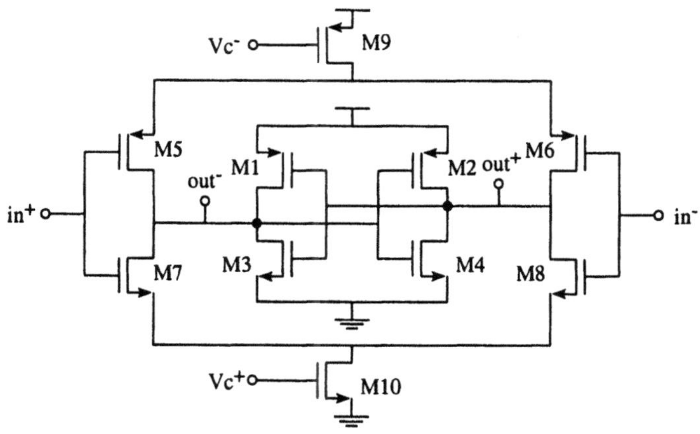

Figure 7.1. Ring oscillator

Table 7.1. Device sizes for the circuit in Figure 7.1

Device(s)	M1 & M2	M3 & M4	M5 & M6	M7 & M8	M9	M10
m	3	1	6	2	6	2
$W(\mu m)$	1.68	1.68	2.40	2.40	2.40	2.40
$L(\mu m)$	0.24	0.24	0.24	0.24	0.24	0.24

The maximum frequency step in Bluetooth is $f_{eo} = 78$MHz, and the frequency divider ratio is approximately 2400. If 100kHz of loop bandwidth is chosen, the PLL lock-in time is approximately $26\mu s$. Therefore, a PLL with an integer-N frequency divider and 100kHz loop bandwidth is sufficient to meet the requirement. Since each channel occupies a 1 MHz bandwidth, the input clock can be chosen to be 1 MHz so that the output frequency is an integer multiple of 1 MHz.

1. PLL with Ring VCO

Based on our ring oscillator that was introduced in Chapter 6, we have completed a PLL design in TSMC $0.25\mu m$ CMOS technology with a fully differential charge pump, loop filter and VCO. The rail-to-rail signal swing in the VCO minimizes the out-of-band phase noise for the PLL. Meanwhile the fully differential topology provides improved power supply and substrate noise rejection.

1.1 Ring VCO

The VCO is a two-stage ring oscillator with the same delay cell topology as the one in Chapter 6. Its delay cell is redrawn in Figure 7.1. The sizes of the devices are listed in Table 7.1.

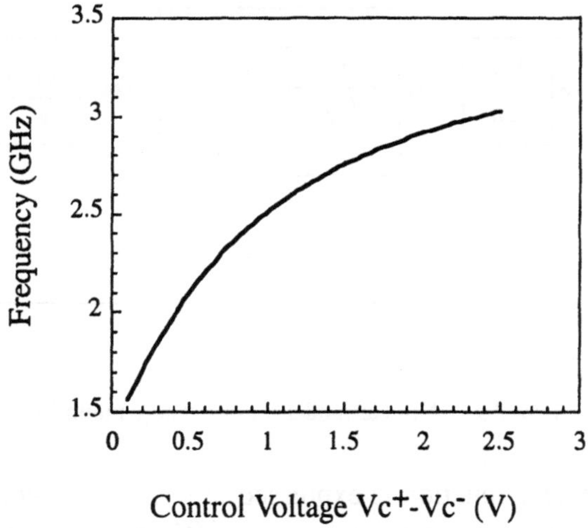

Figure 7.2. Frequency control characteristic for the ring oscillator

Since in our CMOS process, the mobilities for NMOS and PMOS satisfies $\mu_N = 3\mu_P$, all the PMOS transistors in Figure 7.1 are three times as large as their NMOS counterparts. Therefore, the ring oscillator has equal rise and fall times. The symmetry of the waveform helps minimize the influence of device flicker noise [Hajimiri and Lee, 1998]. The equal strength of $M9$ and $M10$ makes the frequency of oscillation insensitive to the common-mode control voltage.

The frequency control characteristic is simulated and plotted in Figure 7.2. The value for $V_c^+ - V_c^-$ is always kept positive so that both $M9$ and $M10$ are in deep triode. The output frequency is centered around 2.4GHz. The tuning range is designed to be large enough to accommodate the process and temperature variation. Figure 7.2 shows the differential tuning characteristic. However, it is easier to calculate the component values for a single-ended PLL, and then simply duplicate the components according to the circuit symmetry and make it differential. In order to suppress the common-mode noise, the ring oscillator is designed such that its frequency has equal sensitivity to V_c^+ and V_c^-. It can be obtained from Figure 7.2 that under nominal operating conditions, the VCO has a single-ended gain given by $K_{VCO} = 8.9 \times 10^9 rad/(sec \cdot V)$.

1.2 Charge Pump and Loop Filter

Our next task is to calculate the resistor and capacitor values in the loop filter and the charge pump current from the loop bandwidth requirement of $K_0 = 2\pi \times 100kHz = 2\pi \times 10^5 rad/sec$.

The diagram in Figure 7.3 shows a linear model for our charge pump PLL, where the PFD gain is expressed in terms of the charge pump current, and

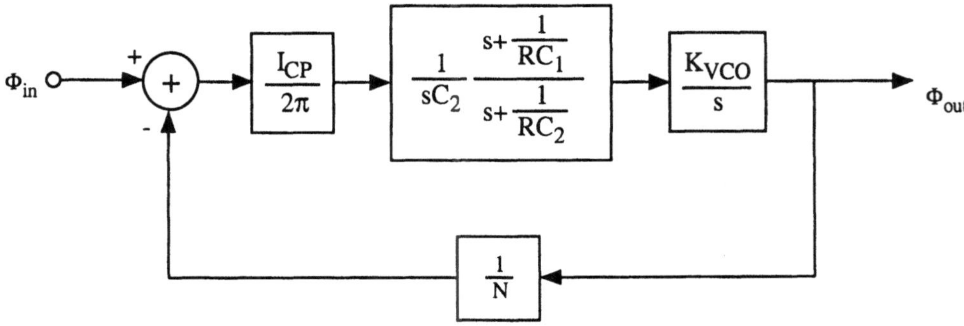

Figure 7.3. A linear model for our charge pump PLL

the loop filter transfer function is expressed in terms of its component values. It consists of an open-loop zero at $\omega_z = 1/(RC_1)$ and an open-loop pole at $\omega_p = 1/(RC_2)$. There is another double pole at DC, i.e., $\omega = 0$.

Recall our discussions in Chapter 2. An optimal choice for ω_z and ω_p is $4\omega_z = K_0 = (1/4)\omega_p$. Now that K_0 is chosen to be $K_0 = 2\pi \times 10^5 \approx 6.28 \times 10^5 rad/sec$, ω_z and ω_p are given by

$$\frac{1}{RC_1} = \omega_z = \frac{1}{4}K_0 \approx 1.57 \times 10^5 rad/sec \tag{7.2}$$

$$\frac{1}{RC_2} = \omega_p = 4K_0 \approx 2.51 \times 10^6 rad/sec \tag{7.3}$$

We have three unknowns but only two equations in Equations 7.2 and 7.3. So we have one degree of freedom in choosing the values for R, C_1 and C_2. From Equation 2.33, it is seen that in order to minimize the noise from the charge pump, $K'_{PD} = \frac{I_{CP}}{2\pi N}$ has to be maximized. Let us rewrite Equation 2.23 here.

$$K_0 = K'_{PD}K_{VCO}R = \frac{I_{CP}K_{VCO}R}{2\pi N} \tag{7.4}$$

It is seen from the above equation that for a given K_0 and K_{VCO}, maximizing K'_{PD} is equivalent to minimizing R. A smaller value for R also reduces the thermal noise from the resistor which is given by $\overline{v_{n,R}^2}/\Delta f = 4kTR$.

Therefore, we would like to minimize the value for R when we solve Equations 7.2 and 7.3. However, a practical difficulty is the limitation of the physical size of an on-chip capacitor. According to Equations 7.2 and 7.3, $C_1 = 16C_2$. Hence we chose C_1 to be maximal size that we can afford on chip, and then determine the values for R and C_2 accordingly.

In the TSMC $0.25\mu m$ CMOS technology, the most area-efficient capacitor is a MOS (either NMOS or PMOS) capacitor. It is the gate capacitance for a MOS transistor when it is turned on. The capacitance is given by $C = C_{ox}WL$,

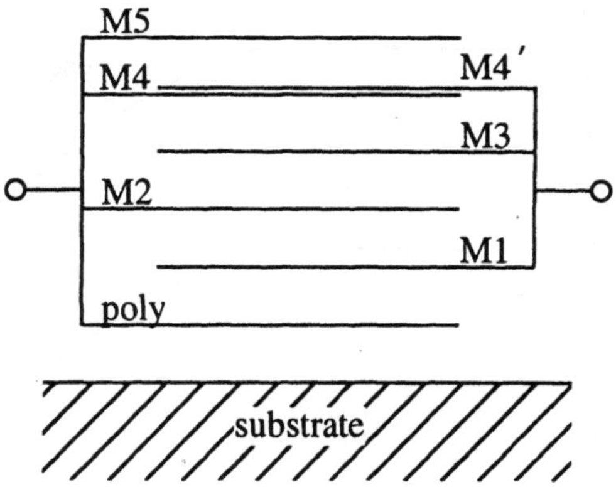

Figure 7.4. A cross-section view of an on-chip capacitor with sandwiched layers

where C_{ox} is the capacitance per unit area which is approximately $6fF/\mu m^2$ for a MOS capacitor in this process. However, when the device is not turned on strongly, its capacitance becomes a function of its gate voltage V_{gs}. Unfortunately, when used in a PLL loop filter, the capacitor may experience large voltage swings which may shut off the MOS transistor temporarily. This turns the capacitor into a voltage-dependent varactor, which is highly undesirable.

Another choice for the capacitors in our technology is MiM capacitors which consist of two metal layers with a thin insulator layer in between. They have much higher quality than the MOS capacitors and their capacitance per unit area is about $0.99fF/\mu m^2$. The CMOS technology that we use has five basic metal layers in total and one polysilicon layer. The bottom plate of the MiM capacitor is the regular Metal 4 layer ($M4$), and the top plate is an optional metal layer ($M4'$)in addition to the basic layers. The vertical spacing distance between $M4$ $M4'$ is much smaller than the spacing between any other layers. When we sandwich all the conductive layers together, we can improve the area-efficiency for our on-chip capacitors to some extent. This is illustrated in Figure 7.4 where the MiM capacitor is implemented between the layers $M4$ and $M4'$. With the sandwiched structure, the area efficiency increases up to about $1fF/\mu m^2$.

Now solving Equations 7.2, 7.3 and 7.4, we obtain the following parameters for the PLL with a divider ratio of $N = 2400$:

$$C_1 = 137.12pF \tag{7.5}$$

$$C_2 = 8.5699pF \tag{7.6}$$

$$R = 46.429k\Omega \tag{7.7}$$

$$I_{CP} = 22.9259\mu A \tag{7.8}$$

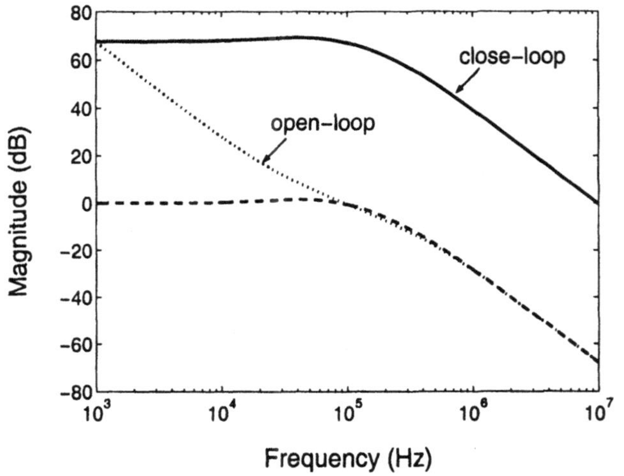

Figure 7.5. Open-loop and Closed-loop transfer functions for the PLL with a ring VCO

The largest capacitor C_1 takes about $0.14mm^2$ of chip area which is affordable in our design. The charge pump current is chosen to be a multiple of our global bias current so that it can be obtained easily by using current mirrors.

The resulting open-loop and closed-loop transfer functions are plotted in Figure 7.5. The dotted curve is the open-loop transfer function which has a 0 dB bandwidth of 100kHz. It has a zero at ω_z = 25kHz and a pole at ω_p = 400kHz. In order to obtain the closed-loop transfer function from the plot, we first obtain the closed-loop transfer function without the divider N. This is plotted in the dashed curve. It stays at 0 dB for $\omega < K_0$ and follows the open-loop curve for $\omega > K_0$. In order to obtain the final closed-loop curve, we simply multiply the existing curve by $N = 2400$ which is equivalent to 67.6 dB. This is shown in the solid curve. It is also seen from this plot that under the condition of $4\omega_z = K_0 = (1/4)\omega_p$, the closed-loop response has a slight peak of 1.6 dB which is good enough for our application.

Figure 7.6 shows its response to input frequency steps. It is simulated with a linear charge pump PLL model. When the input frequency takes a unit step, the output responds by a step of $N = 2400$. It overshoots by 16.9%. In Bluetooth, the largest frequency step is no more than ±78MHz. The frequency synthesizer is required to settle to the desired channel within ±70kHz in $220\mu s$. By doing proper scaling, Figure 7.6 indicates that our PLL settles to the required precision within $29\mu s$. In a practical case, even considering possible cycle slips and other nonlinear capture processes, our design will still meet the Bluetooth specification for settling time.

In order to achieve good supply and substrate noise rejection, both the charge pump and the loop filter have to be fully differential. We take the circuit topology

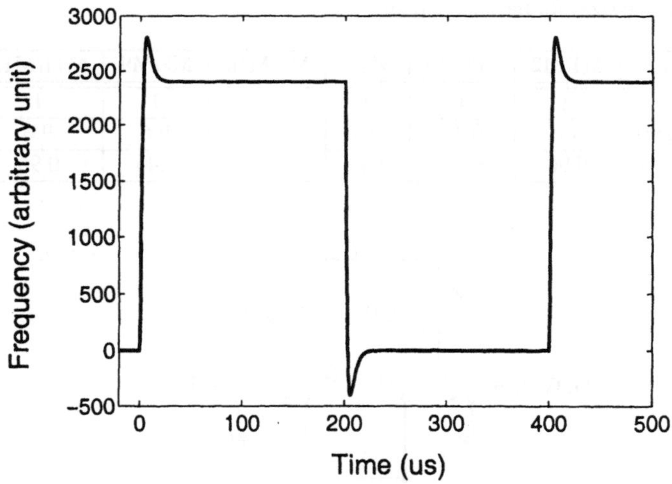

Figure 7.6. Frequency step response for the charge pump PLL

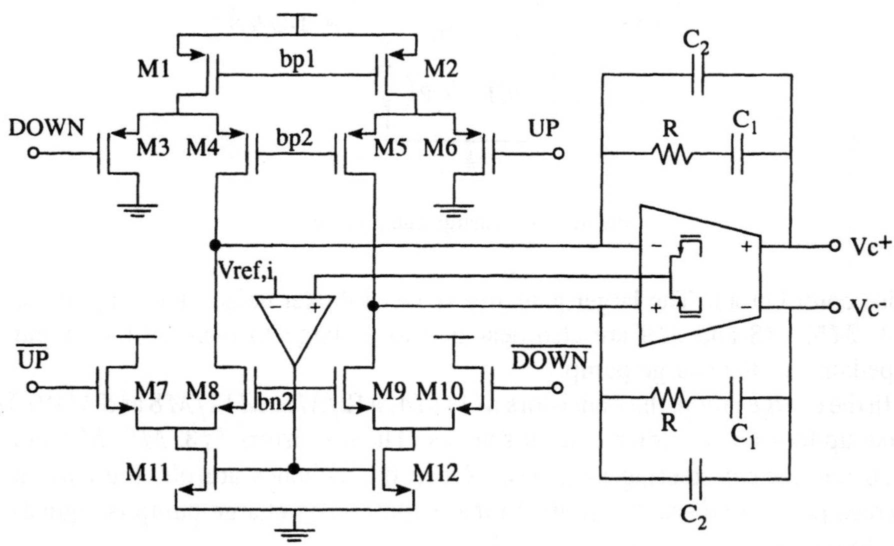

Figure 7.7. Differential charge pump and loop filter

in [Li et al., 2000] and adopt it into our design. The charge pump and the loop filter is shown in Figure 7.7, and the device sizes are listed in Table 7.2.

As we have discussed in Chapter 2, the noise from the charge pump is low-pass filtered. Therefore, the low-frequency noise, in particular, the flicker noise from the charge pump current, has to be small enough such that the close-in phase noise satisfies Bluetooth, requirement. So we avoid using smallest channel lengths for transistors $M1$, $M2$, $M11$ and $M12$, which are the major

Table 7.2. Device sizes for the charge pump

Devices	M1/M2	M3/M6	M4/5	M7/M10	M8/M9	M11/M12
m	3	1	3	1	1	1
W (μm)	9.6	0.48	9.6	0.48	9.6	9.6
L (μm)	0.96	0.48	1.44	0.96	1.44	0.96

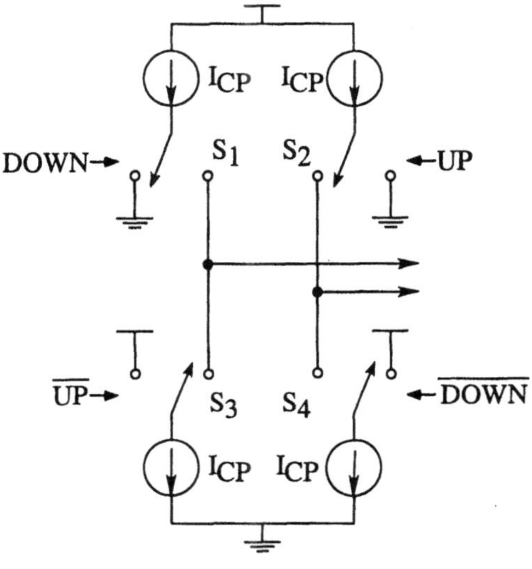

Figure 7.8. Charge pump model

noise contributors. The larger gate area reduces flicker noise. The lengths for $M4$, $M5$, $M8$ and $M9$ are also designed to be large to increase the output impedance of the charge pump.

In the charge pump, the transistors $M1/M4$, $M2/M5$, $M11/M8$ and $M12/M9$ make up four up and down current sources. The transistors $M3$, $M6$, $M7$ and $M10$ are current steering switches. When the switches are off, the current sources provide constant currents to the outputs. The charge pump is equivalent to the model drawn in Figure 7.8.

This circuit has several advantages compared to the conventional charge pump included in Figure 2.6. Since the gates for the transistors $M4$, $M5$, $M8$ and $M9$ are biased at constant voltages, the output nodes do not suffer from clock feedthrough. Also since they are working in saturation before they are turned off, the channel charge injection does not affect the output nodes [Dai and Harjani, 2000b].

The transistors $M3$, $M6$, $M7$ and $M10$ are designed to be weak transistors that have significant channel resistance when they are on. For example, when $M7$ is turned on, it only pulls its source voltage up by a few hundred millivolts so that $M8$ is turned off. This has to be check over temperature and process variation to ensure its correct functionality. Since the parasitic capacitance at the drain of $M11$ does not have to charged and discharged completely, and $M11$ remains in saturation all the time, it makes the current steering very fast and reduces possible clock skew.

Very small transistors are used for $M3$, $M6$, $M7$ and $M10$. This makes the parasitic capacitances due to their source pn junction negligible. Charging and discharging those parasitic capacitances is equivalent to changing the effective charge pump current instantaneously. As we have discussed before, mismatch in the charge pump currents results in reference spur at the PLL output spectrum. Therefore, small switch devices and small voltage swings at the sources of $M3$, $M6$, $M7$ and $M10$ minimize the reference spur. Meanwhile, the clock feedthrough due to their C_{gs} actually helps increase the switching speed, because it increases $|V_{gs}|$ for transistors $M4$, $M5$, $M8$ and $M9$ instantaneously and helps them turn on.

Compared to the single-ended version in Figure 2.6, the differential version has significant improvement in current matching. Instead of trying to match the currents through the PMOS current mirror and the NMOS current mirror, now we try to match the left side of the circuit with the right side of the circuit. Since the drain voltages for the transistors $M4$, $M5$, $M8$ and $M9$ are approximately equal due to the virtual ground provided by the opamp in the loop filter, the mismatch between the current mirrors is minimized and the finite output impedance of the current mirrors becomes unimportant. Even the mismatch between the PMOS and NMOS currents still exists, it only disturbs the common mode, which is suppressed by the common-mode feedback of the opamp in the loop filter and the common-mode rejection of the differential VCO.

The common-mode voltage at the inputs of the opamp needs to be well defined. This is controlled by a common-mode feedback (CMFB) circuit shown in Figure 7.7. It is realized with an amplifier whose transistor-level schematic is drawn in Figure 7.9, and whose devices sizes are listed in Table 7.3. It is intentionally designed as a transconductance amplifier with low voltage gain in order to avoid common-mode instability. Large loop gain is achieved with the help of the large output impedance of the charge pump. The non-inverting input of the transconductance amplifier is connected to the common source of the differential pair in the opamp in the loop filter (node c as later shown in Figure 7.10). Since the two inputs of the opamp are held at approximately equal voltages, their common-source voltage is their common-mode voltage downshifted by V_{gs} for the differential pair. Now this voltage is compared with $V_{ref,i}$ in Figure 7.7. The amplifier converts the common-mode error voltage into a current, which is transferred into a voltage by the current mirrors in the charge pump. Therefore the common-mode at the input of the loop filter is well controlled with a very small error.

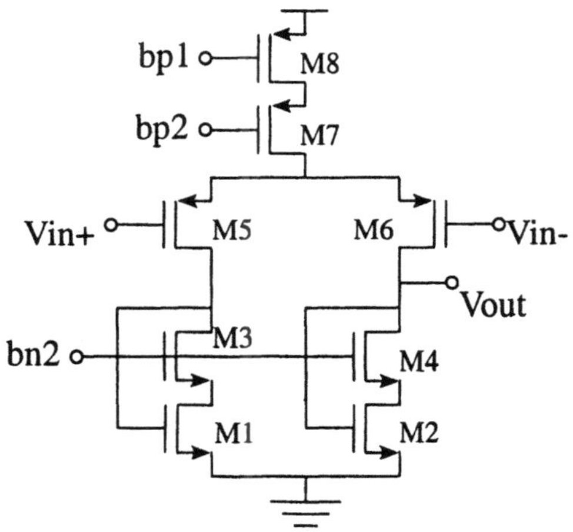

Figure 7.9. Amplifier for CMFB

Table 7.3. Device sizes for the amplifier used in the CMFB

Devices	M1/M2	M3/M4	M5/M6	M7	M8
m	1	1	24	5	5
W (μm)	9.6	9.6	9.6	9.6	9.6 &
L (μm)	0.96	1.44	0.96	1.44	0.96

It is worth noticing that the CMFB in Figure 7.7 is active only when both $\overline{UP}$ and $\overline{DOWN}$ are low. Otherwise the control loop is not closed. In a dead zone free PLL, this is true only for a very short period of time when the PLL is phase-locked. Both UP and $DOWN$ signals stay low during most of the clock period. They are turned high long enough only to remove the dead zone. Therefore, the CMFB works in discrete time. Once UP or $DOWN$ are turned high, the common-mode is adjusted, and once they are turned low, the feedback loop is broken, and the common-mode may drift. This can possibly generate some common-mode ripple. Fortunately, because of the outputs of the charge pump are kept at high impedance when the CMFB loop is broken, the common-mode drift is minimum and this is suppressed further by the common-mode rejection of the opamp in the loop filter and the VCO with differential frequency tuning.

The opamp used in the loop filter is a fully differential folded cascode opamp as shown in Figure 7.10. It consists of a gain stage followed by source followers. Since $V_{out}^{+} - V_{out}^{-}$ always has to be kept positive, which is required by our ring oscillator, the source followers at the opamp outputs are used for voltage shifting. The common-mode voltage of the gain stage outputs at nodes a and b are measure by a voltage divider consisting of two $200k\Omega$ resistors. We have

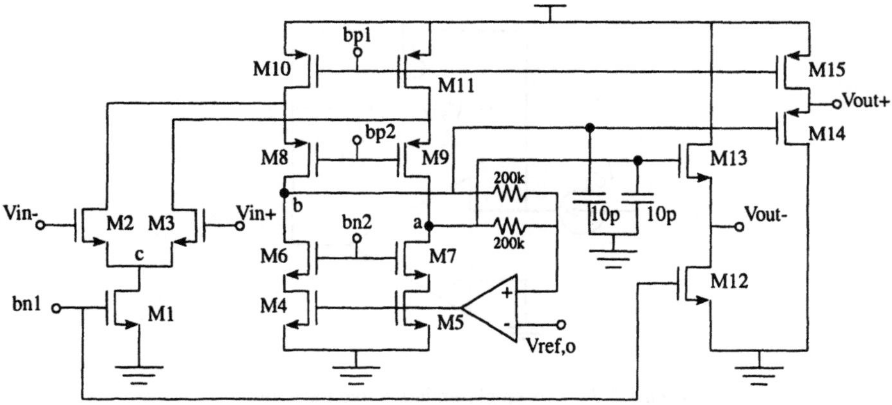

Figure 7.10. Folded cascode opamp with CMFB and source followers

Table 7.4. Device sizes for the folded cascode opamp

Devices	M1	M2/3	M4/5	M6/7	M8/9	M10/11	M12	M13	M14	M15
m	4	32	2	2	6	12	20	160	480	60
W (μm)	9.6	9.6	9.6	9.6	9.6	9.6	9.6	2.4	2.4	9.6
L (μm)	0.96	0.96	0.96	1.44	1.44	0.96	0.96	0.24	0.24	0.96

chosen the resistance to be large enough so that the gain for the opamp is not significantly reduced. The CMOS technology that we have chosen provides lightly-doped polysilicon which can be used as resistors, and $200k\Omega$ can be implemented with affordable chip area.

The common-mode voltage taken between the two resistors is compared with the reference voltage $V_{ref,o}$ with the same amplifier as shown in Figure 7.9. The reason that the common-mode voltage is measured at nodes a and b instead of V_{out}^{+} and V_{out}^{-} is because the voltages at a and b are much closer to each other than nodes V_{out}^{+} and V_{out}^{-}, and the resistors will draw less current if they are connected between a and b. This CMFB circuit works in continuous time as opposed to the input CMFB circuit which works in discrete time. Therefore, any common-mode ripple at the opamp inputs can be effectively rejected. Further, two $10pF$ capacitors are connected at a and b to ensure stability.

Despite the advantages of good common-mode rejection and availability of virtual grounds at the inputs by using an active loop filter, there is also a disadvantage of it compared to using a passive filter. An active filter involving an opamp is usually much noisier than a passive filter which only consists of resistors and capacitors. In particular, the flicker noise due to the MOS transistors is likely to dominate the low frequency phase noise. We have to pay special attention to make sure that flicker noise does not cause serious problems.

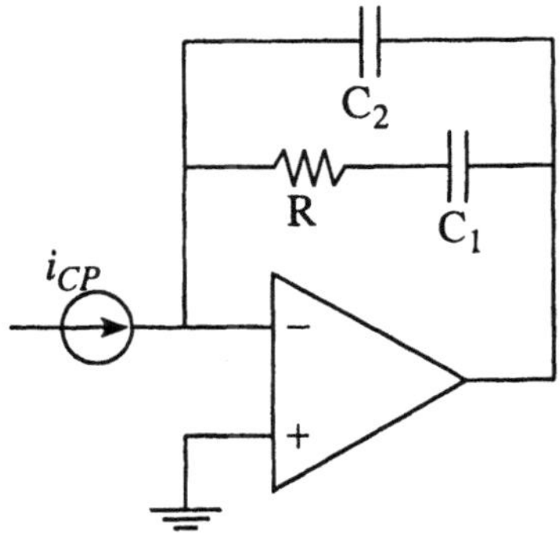

Figure 7.11. A single-ended active loop filter

For frequency synthesizers, the out-of-band phase noise is more critical. In our design, the loop bandwidth is 100kHz. Comparing with a passive filter with the same resistor and capacitor values, we would like to have the flicker noise level not to exceed the thermal noise level due to the passive resistor R for frequencies $f > 100$kHz. If this is achieved, the out-of-band phase noise is not degraded dramatically by the flicker noise due to the opamp.

For simplicity, we have drawn a single-ended active loop filter in Figure 7.11. The analysis for the fully differential filter can be implied accordingly. Since the flicker noise has a $1/f$ PSD and the thermal noise has a white PSD, if we make sure that the level for the flicker noise at 100kHz does not exceed the thermal noise, we know that it will not make a major contribution for $f > 100$kHz since its PSD drops at a faster rate than the thermal noise.

As we have analyzed earlier, at 100kHz the impedance of the loop filter can be approximated by R. Assuming the charge pump has a infinite output impedance, the filter in Figure 7.11 can be redrawn in Figure 7.12. Also shown in the figure are the thermal noise source due to the resistor and the flicker noise source due to the opamp. They are uncorrelated noise sources. Since the opamp has to be designed with a significantly larger bandwidth than the loop bandwidth, its gain can be approximated by infinity at 100kHz. Therefore, the total output noise power is simply the sum of the power of the two noise sources.

We have derived earlier in this chapter the resistor value to be $46.429k\Omega$. At normal room temperature of 27°, its thermal noise is given by

$$\overline{v_{R,n}^2}/\Delta f = 4kTR = 7.69 \times 10^{-16} V^2/Hz \tag{7.9}$$

The input-referred flicker noise is calculated by Equation 7.10 in the HSpice Level 3 Model [Meta-Software, 1996]. Since NMOS transistors have much larger flicker noise than PMOS transistors with the same dimension, we are more concerned with $M2$, $M3$, $M4$ and $M5$ in Figure 7.10. The noise due to $M1$ and the amplifier used in CMFB only affects the common-mode and does

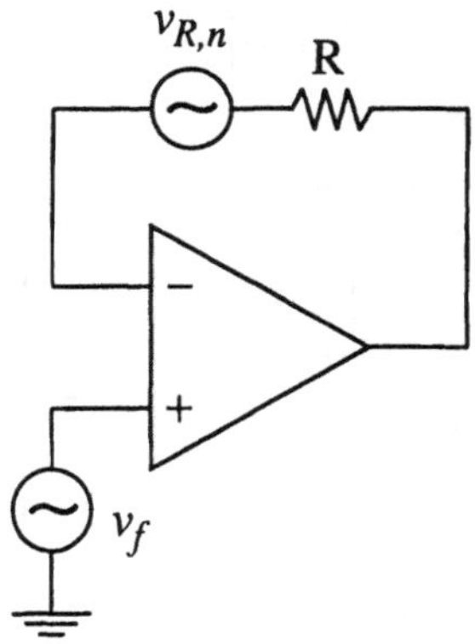

Figure 7.12. Loop filter noise model at 100kHz

not contribute to the differential output signal. The noise due to $M6$ and $M7$ is suppress by the cascode devices. Also the noise due to the source followers are reduced by the gain of the preceding stage.

$$\frac{\overline{v_{R,n}^2}}{\Delta f} = \frac{K_F}{C_{ox}W_{eff}L_{eff}f^{A_F}} \tag{7.10}$$

After careful calculation, we have designed $M2$ and $M3$ to be $32 \times 9.6\mu m \times 0.96\mu m (m \times W \times L)$ and $M4$ and $M5$ to be $2 \times 9.6\mu m \times 0.96\mu m$. Our simulation shows $-142.45dBV/Hz$ output-referred noise at 100kHz due to flicker noise. Considering the VCO gain, this transfers into $-71.73dBc/Hz$ PLL phase noise at 100kHz offset frequency.

1.3 Phase/Frequency Detector

Our PFD design is based on the prototype introduced in Chapter 2. It generates UP and $DOWN$ signals with equal duration for a short period of time in each clock cycle in order to eliminate dead zone. It also has to provide $\overline{UP}$ and $\overline{DOWN}$ signals which are required by the fully differential charge pump.

Figure 7.13 shows our design for the PFD. It is a modified version of the design in Reference [Maneatis, 1996]. A few inverters have been added in order to eliminate hazard conditions. Also they make the duration for the $\overline{UP}$ and $\overline{DOWN}$ pulses long enough so that no dead zone exists. This design is one of the possible implementations for the prototype in Figure 2.5 in Chapter 2.

The design of PFD in Figure 7.13 only provides $\overline{UP}$ and $\overline{DOWN}$. We still need to generate UP and $DOWN$ signals. There should be no significant delay or skew between all these signals. Otherwise, there could be an increase in the common-mode ripple or even differential ripple on the VCO control signal. In order to align them precisely, we have designed the circuit in Figure 7.14 to

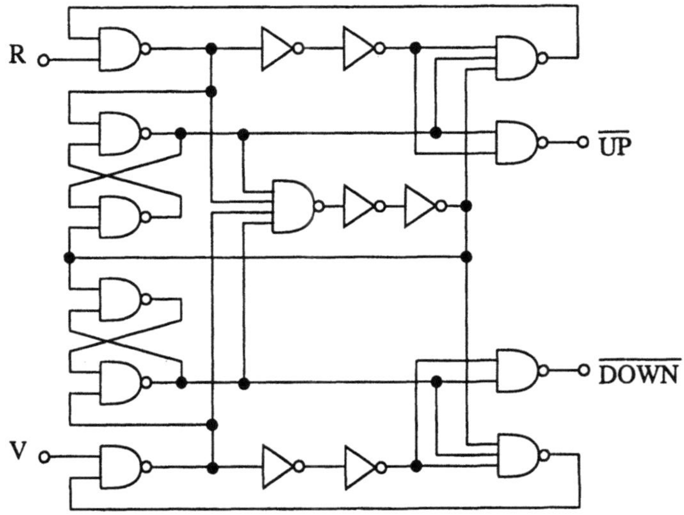

Figure 7.13. PFD with no dead zone

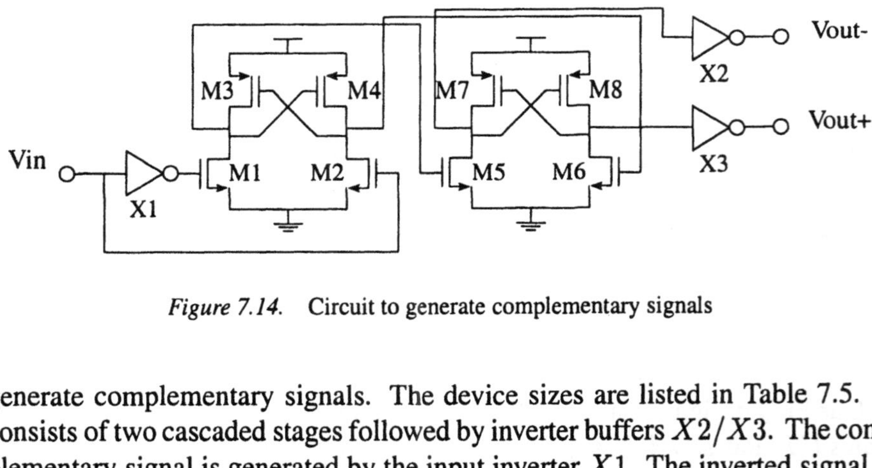

Figure 7.14. Circuit to generate complementary signals

generate complementary signals. The device sizes are listed in Table 7.5. It consists of two cascaded stages followed by inverter buffers $X2/X3$. The complementary signal is generated by the input inverter $X1$. The inverted signal is delayed from the original input signal because of the inverter. Due to the positive feedback by the cross coupled PMOS transistors $M3/M4$ and $M7/M8$, the transitions on both sides of the outputs are forced to occurs at the same time. Thus the skew between the complementary outputs is minimized. This is improved even further by cascading two such stages. To ensure robustness, the outputs are buffered by two inverters $X2$ and $X3$.

1.4 Prescaler and Frequency Divider

Due to the high-speed operation of the VCO, the design of the feedback frequency divider in CMOS is also a challenging job. Actually it is the major limitation of high-speed PLL designs in CMOS. It can consume a lot of power

Table 7.5. Device sizes for the complementary signal generator

Devices	M1/M2/M5/M6	M3/M4/M7/M8
m	1	1
W (μm)	1.92	2.88
L (μm)	0.24	0.24

and generate a large amount of jitter. In most cases, static CMOS logic circuits are not fast enough for the first few stages in the frequency divider where speed is most critical. Therefore, a special type of divider is introduced, which is called a *prescaler*.

As we have discussed in Chapter 2, the jitter generated in the feedback divider is low-pass filtered as it is transferred into the PLL output jitter. Therefore, it may add in-band noise if its phase noise is large. In extremely high-speed applications the circuits can barely make a complete transition. The devices work the active region most of the time and act as amplifiers. As the signal propagates through the divider, a lot of circuit noise is added in terms of jitter. Furthermore, most of the divider designs are single-ended and they are susceptible to supply and substrate noise. Therefore, high-speed and low-noise prescaler designs are hard problems.

Another consideration in a feedback frequency divider design is its programmability. For a divider ratio of N, the input frequency and the output frequency obey the relationship of $f_{out} = N f_{in}$. The input of the PLL is usually taken from a crystal oscillator which has a fixed frequency f_{in}. The PLL output frequency can be digitally controlled by changing the divider ratio N. For example, in the US, Europe and most other countries, the Bluetooth specification allocates frequencies (2402+k) MHz for its RF channels, where $k = 0, ..., 78$. Therefore, if a 1 MHz reference clock is taken from a crystal oscillator, the feedback divider ratio has to be programmed ranging from 2402 to 2480.

A well-known implementation of such a programmable frequency divider is called *pulse swallow frequency divider* which is shown in Figure 7.15 [Razavi, 1998]. It consists of a prescaler, a program counter and a swallow counter. The prescaler divides the input either by N or $N+1$ depending on the logic level of the modulus control. The program counter always divides the prescaler output by P. The swallow counter divides the prescaler output by S where S varies from 1 to the maximum number of channels, and it is determined by the channel selection. P is usually chosen to be larger than S.

The pulse swallow frequency divider works as follows. As a reset state, the prescaler divides by $N+1$. After $S(N+1)$ cycles at the input, the output of the divide-by-S swallow counter changes its output logic level, making the prescaler divide by N. By now, the divide-by-P program counter has seen S

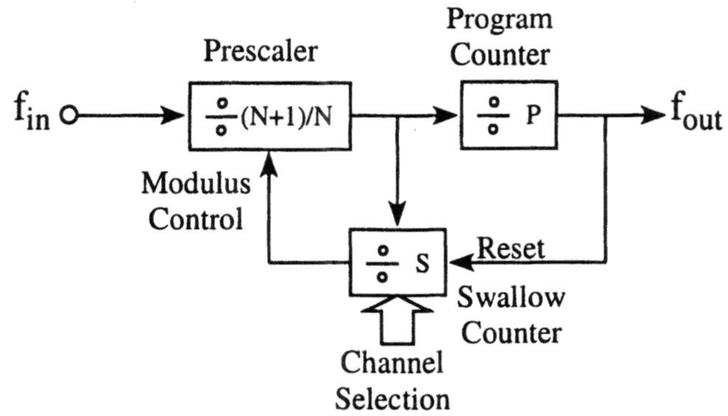

Figure 7.15. Pulse swallow frequency divider

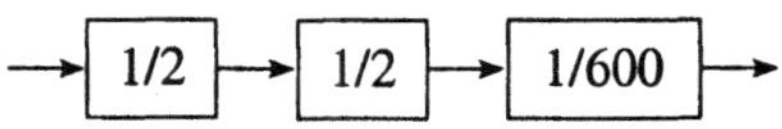

Figure 7.16. Top level diagram for feedback divider

cycles at its input. Hence, it needs $P - S$ more cycles to change its output state. This corresponds to $(P - S)N$ cycles at the main input. After the $(P - S)N$ more cycles at the input, the program counter has completed a cycle at its output and resets the swallow counter. This also switch the prescaler back to the divide-by-$N + 1$ mode. Therefore, in order to obtain a complete cycle at the output, a total of $S(N + 1) + (P - S)N = PN + S$ cycles at the input are needed. By changing S, the divider ratio can be easily programmed. In the case of a frequency synthesizer for Bluetooth, PN can be chosen as 2400 and S can be variable between 2 and 80.

Due to the time constraint and the maturity of digital design techniques, we did not spend time designing the fully programmable feedback divider. Instead, we designed a divider with a fixed ratio of 2400. This does not jeopardize our research goal of low-phase-noise design.

Our frequency divider circuit consists of two divide-by-2 stages and one divide-by-600 stage as shown in Figure 7.16. Since our input frequency is 2.4GHz, static CMOS logic circuits cannot be used for the divide-by-2 due to the speed limitation. After dividing the frequency down by four, a CMOS logic circuit is used for the divide-by-600 in order to improve robustness.

The first two divide-by-2 stages are identical and the circuit schematic is shown in Figure 7.17 [Wang, 2000]. It is based on two D-latches that can be either in the flipping mode or the latched mode. In the flipping mode, the tail device $M1$ or $M8$ is on and the PMOS devices $M6/M7$ or $M13/M14$ are also on. The PMOS devices have low impedance and can charge up the parasitic capacitance at one of the output nodes quickly. The voltage at the other output

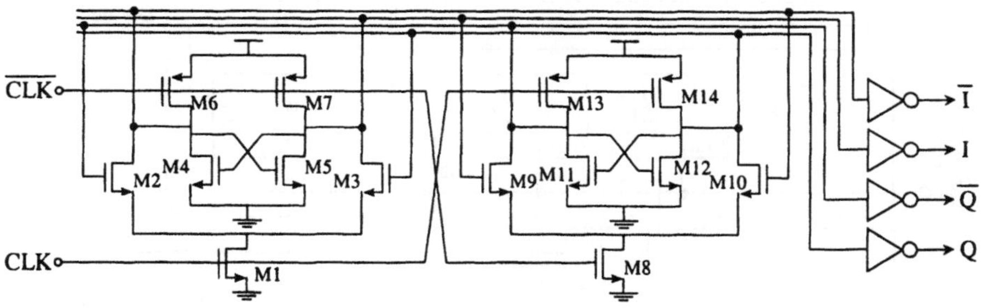

Figure 7.17. Divide-by-2 circuit

Table 7.6. Device sizes for prescaler

Devices	M1/M8	M2/M3/M9/M10	M4/M5/M11/M12	M6/M7/M13/M14
m	4	2	2	2
W (μm)	2.4	2.4	2.4	2.4
L (μm)	0.24	0.24	0.24	0.24

node also goes down quickly since the NMOS transistors are much stronger than the PMOS transistors. In the latched mode, both the tail devices and the PMOS devices are off. One of the output nodes are connected to the ground through the NMOS transistor, and the other output node is held at high impedance. The up swing of $\overline{CLK}$ helps keep the voltage level high, since it charges up the output node via the parasitic capacitance through C_{gd} of the PMOS transistors. This circuit automatically generate quadrature outputs include I, $\overline{I}$, Q and $\overline{Q}$. In order maintain the symmetry of the circuit regardless of the loading, four digital inverters are used as output buffers to provide isolation from the loads. In our design, we do not need quadrature outputs from the prescaler and we only take I and $\overline{I}$ as the outputs. The device sizes are listed in Table 7.6.

The divide-by-600 circuit is shown in Figure 7.18. It consists of a dual-modulus divide-by-4/5 divider and a divide-by-128 divider. The dual-modulus divider divides by 4 when the $cntrl$ signal is low, and it divides by 5 when the $cntrl$ signal is high. In the initial state when the output is high, the $cntrl$ is low, and the total divider ratio is $4 \times 128 = 512$. After counting 40 cycles at the divide-by-128 input, i.e., 160 cycles at the main input, the feedback combination logic sets the $cntrl$ signal high so that the total divider ratio is now $5 \times 128 = 640$. After 88 more cycles at the divide-by-128 input, the main output toggles back to high again, since it has counted a total of $40 + 88 = 128$ cycles. This corresponds to a total of $40 \times 4 + 88 \times 5 = 600$ cycles at the main input. Now the $cntrl$ signal is set low and the first stage starts to divide by 4, which is its initial state. Therefore, the complete divider has an average ratio of 600.

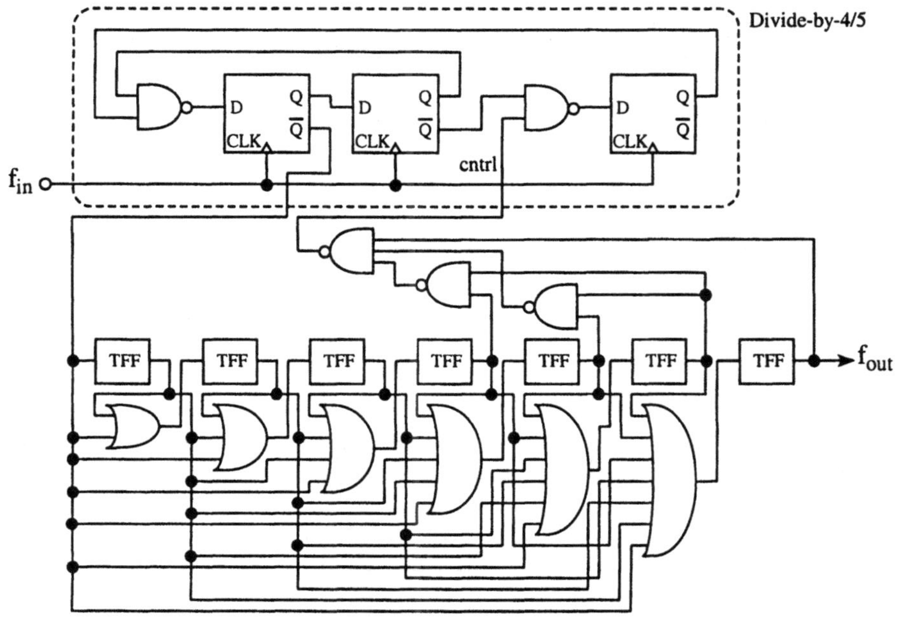

Figure 7.18. Divide-by-600 circuit

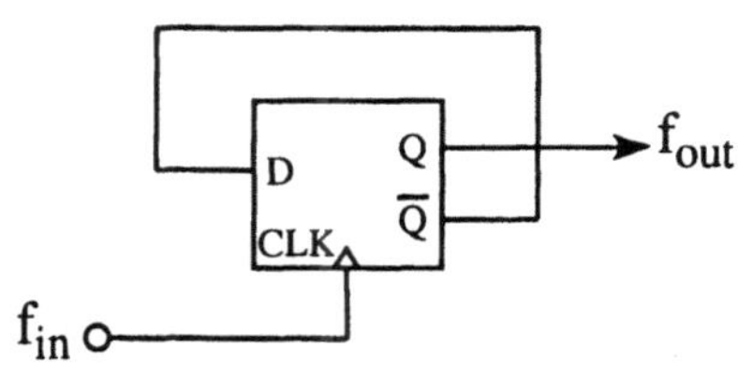

Figure 7.19. A divide-by-2 toggle flip-flop

The divide-by-128 circuit is implemented by cascading 7 toggle flip-flops, each of which is a divide-by-2 circuit as shown in Figure 7.19. The D-flip-flops in Figures 7.18 and 7.19 are implemented by the circuit shown in Figure 7.20.

The D-flip-flop in Figure 7.20 is a dynamic CMOS logic, which contains fewer number of transistors than a static CMOS D-flip-flop. This results in lower power consumption and faster speed. It consists of a P-latch, an N-latch and an inverter. Both the P-latch and the N-latch are realized by using two clocked inverters. The transistors Mn and Mp are very weak transistors that are used to eliminate the high-impedance state and improve the circuit's robustness.

When 7 toggle flip-flops are cascaded, each stage contributes some phase uncertainty, or jitter. The total jitter in the output clock is the sum of the jitter contributed by each stage. As we have discussed in Chapter 2, the phase

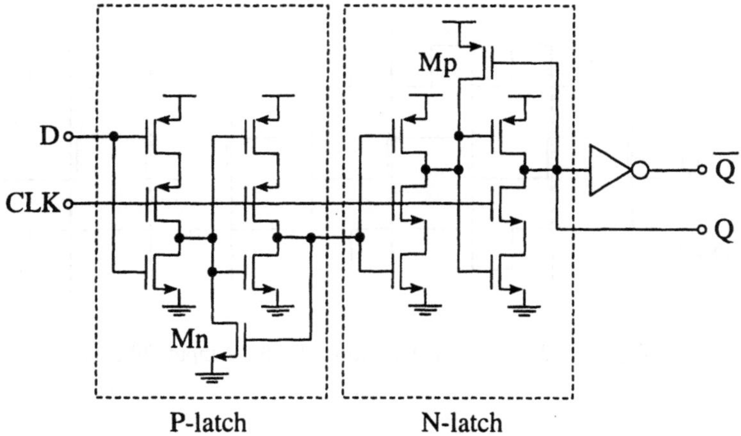

Figure 7.20. A D-type flip-flop

noise generated by the feedback frequency divider is low-pass filtered and may contribute to the in-band noise. In order to reduce the noise generated by the toggle flip-flops, we have used the *OR* gates in Figure 7.18 [Egan, 1990]. In order to illustrate how the *OR* gates reduces the total output jitter, we have drawn an example waveforms for the 2-input *OR* gate in Figure 7.21. The frequency for Q is the frequency for D divided by 2. Furthermore, its transition edges are delayed slightly from the edges of D due to the delay of the toggle flip-flop. It also adds some jitter because of the circuit noise. The signal $D + Q$, which is used to trigger the next toggle flip-flop, is also drawn in the figure. Its rising edges are completely determined by the rising edges of the D signal. Hence its output jitter is the sum of the jitter in D and the jitter due to the *OR* gate. The same argument is valid for all the *OR* gates in Figure 7.18. In particular, the jitter in the divide-by-128 is only determined by the jitter at its input and the additional jitter due to the 7-input *OR* gate whose schematic is shown in Figure 7.22. It is implemented by cascading only two stages. Compared with the straightforward implementation by cascading 7 toggle flip-flops, each of which has five stages as shown in Figure 7.20, the amount of additional jitter is significantly reduced.

We have described each building block for our PLL design. The loop parameters have been optimized to minimize phase noise and ensure loop stability. Fully differential topologies have been adopted for the charge pump, loop filter and VCO so that the influence of the supply and substrate noise is minimized.

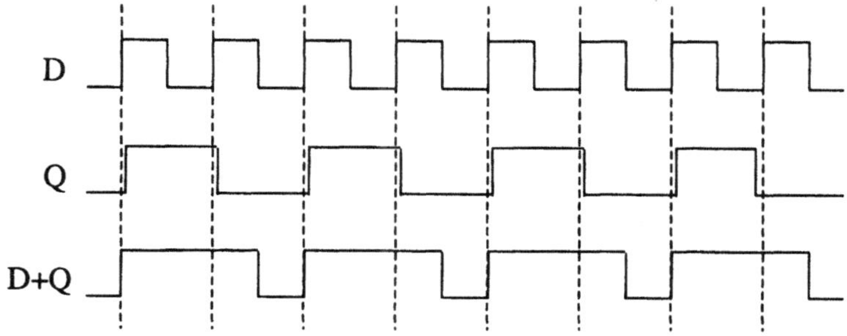

Figure 7.21. Example waveforms for the 2-input *OR* gate

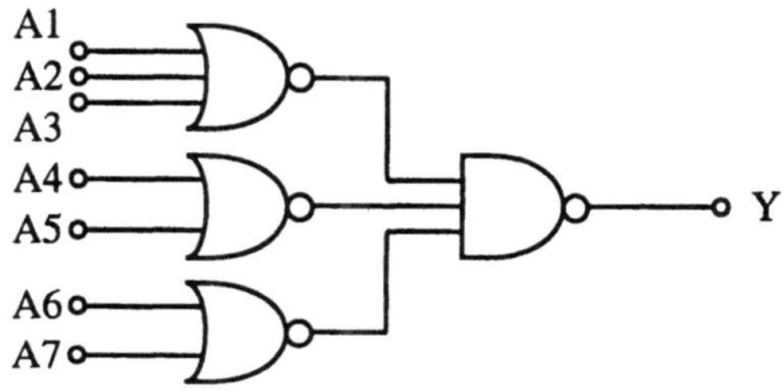

Figure 7.22. Schematic for the 7-input *OR* gate

2. LC VCO

We have analyzed the PLL with a ring oscillator including each of its building blocks. Now we would like to introduce its counterpart that we have designed with an LC VCO.

The circuit schematic for the LC VCO is shown in Figure 7.23. The PMOS transistor $M1$ provides a static current to the oscillator tank. The cross coupled NMOS transistors $M4/M5$ and PMOS transistors $M2/M3$ are equivalent to negative resistance which compensates for the energy loss in the LC tank.

What is unique for this LC VCO circuit is the differential frequency tuning. A total of four varactors are used for frequency tuning instead of two varactors as in a traditional design. By using varactors with the opposite polarities, the common-mode voltage for V_c^+ and V_c^- are cancelled out.

The varactors in Figure 7.23 are accumulation mode MOS varactors [Soorapanth et al., 1998, Porret et al., 2000]. The cross-section view of an accumulation mode MOS varactor is shown in Figure 7.24. It is an NMOS device built in an N-well. The poly gate is n-type doped during the source/drain ion implantation. This causes the area underneath the gate to be in partial depletion for zero V_{gs}. As the gate voltage increases, the device enters more and more into the accumulation region. Now the silicon surface becomes highly conductive and the gate capacitance approaches its maximum value. As the gate

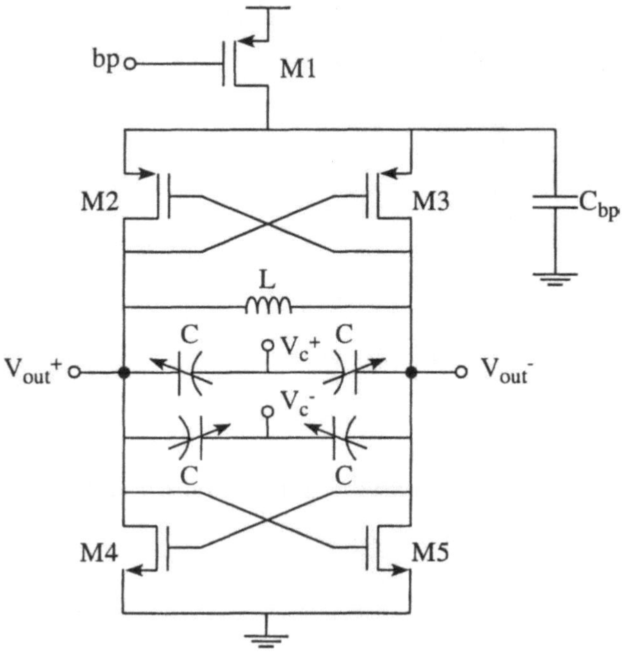

Figure 7.23. Schematic for the LC VCO

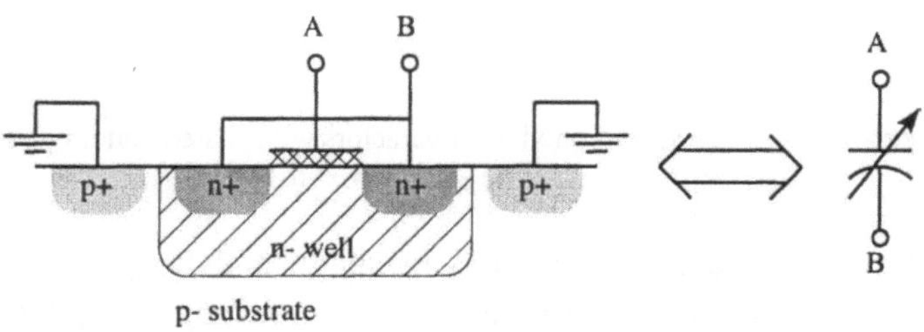

Figure 7.24. Cross-section view of an accumulation mode MOS varactor

voltage decreases, the depletion region widens and the capacitance approaches its minimum value. The C-V characteristic for an accumulation mode MOS varactor is drawn in Figure 7.25. In order to maximize its Q factor, each of the varactors in Figure 7.23 is implemented with multiple MOS devices with $W = 9.6\mu m$ and $L = 0.96\mu m$. Each of the top two varactors is comprised of 12 unit devices and each of the bottom varactors is comprised of 13 unit devices. Based on our simulations, such a choice of capacitance ratio between the top and bottom varactors cancels out the common-mode control voltage under nominal operating conditions.

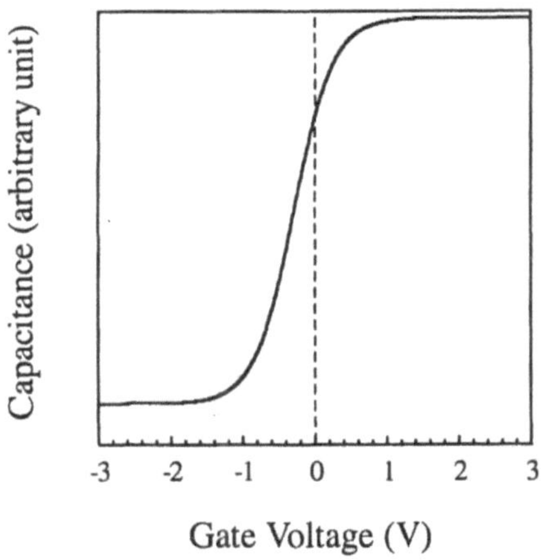

Figure 7.25. C-V characteristic for an accumulation mode NMOS varactor

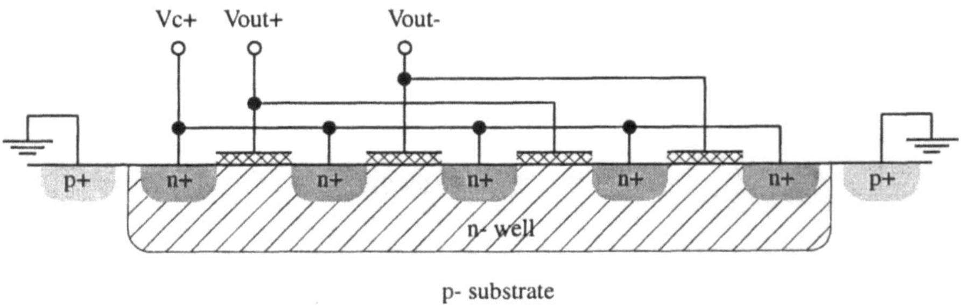

Figure 7.26. A cross-section view of varactors with shared bottom plates

For the top two varactors, the N-wells are shared and the gates with opposite polarities are inter-digitized in order to improve matching and Q factor [Porret et al., 2000]. A cross-section view of the layout is shown in Figure 7.26. For the bottom two varactors, it is hard to inter-digitize the unit devices with opposite polarities, since their gates are connected together and their N-wells are separated. A cross-section view of the layout is shown in Figure 7.27. Because of the separated N-wells, they take larger area than the two varactors at the top.

The N-well and the p-type substrate form a reverse biased p-n junction. One may be concerned with the parasitic capacitance between the N-well and the substrate when the bottom plates of the varactors are connected to the VCO output. It is modelled as shown in Figure 7.28. It may potentially have two impacts on the VCO performance. First, it changes the total capacitance in the tank by adding C'. Second, it reduces the Q factor by adding the substrate and N-well resistance R'.

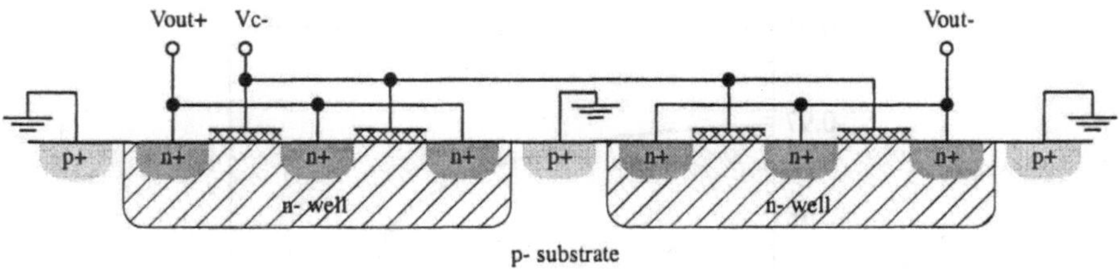

Figure 7.27. A cross-section view of varactors with shared top plates

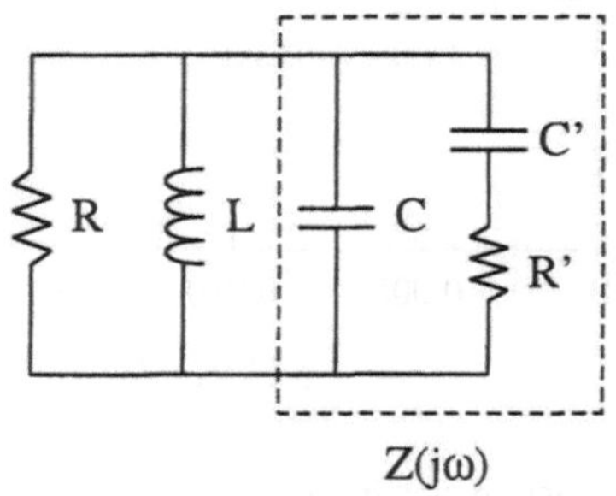

Figure 7.28. A model for an RLC tank with parasitic capacitance between the N-well and substrate

According to the wafer testing results provided by MOSIS, the N-well-to-substrate capacitance is only about 1% of the MOS gate capacitance for the same area. Considering the overhead area for the N-well due to the area needed for the device source, drain and the N-well contacts, and also the voltage dependency of the capacitance, the capacitance between the N-well and the substrate should be smaller than 3% of the NMOS gate capacitance. In other words, $C' < 3\%C$ in Figure 7.28. Therefore, in our design, the varactors have to be made slightly smaller to keep the same capacitance value. This reduces the frequency tuning range slightly, but the impact is limited due to the small N-well-to-substrate capacitance.

We have also studied its impact on the Q factor. Assuming that $C' = 0.03C$, we have calculated $Z(\omega)$ in Figure 7.28. The real and imaginary parts of $Z(j\omega)$ for different R values are plotted in Figure 7.29. The exact N-well and substrate resistance is unknown. However, based on the one of the definitions for Q factor, $Q = |Im(Z)/Re(Z)|$, the worst Q factor for the overall effective varactor is about 68 according to Figure 7.29. Since this is far better than the Q factor achievable for the on-chip spiral inductors, the reduction of Q factor due to the N-well is not a limiting factor for the phase noise.

We have paid special attention to the symmetry of our control terminals. In particular, we would like to keep the loading on both lines approximately the same. For the top two varactors, the control line is connected to the gate

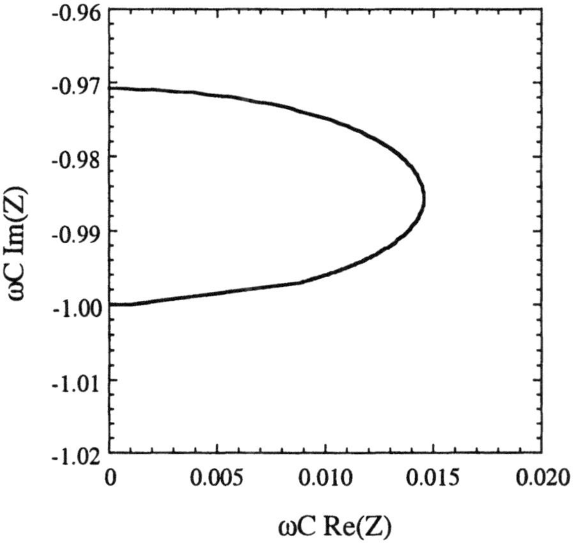

Figure 7.29. Real and imaginary parts of $Z(j\omega)$ for different R values

of the NMOS transistor. Therefore, there is little parasitic capacitance to the substrate. For the bottom two varactors, the control line is connected to the N-well. So there is additional parasitic capacitance to the substrate. Even though this additional parasitic capacitance is small as we have mentioned before, we would like to improve the symmetry of the circuit by connecting the V_c^+ line in Figure 7.23 to an N-well of the same size as the N-well for the bottom two varactors. Its capacitance to the substrate keeps the loading on both control lines approximately the same and minimizes the common-mode distortion. Since the substrate noise picked up by the N-wells is about the same for both control terminals, it only affects its common-mode and is rejected by the VCO.

A critical component that determines the phase noise for an LC VCO is the on-chip spiral inductor. Typically the Q factor for a spiral inductor is below 10. More often, it is below 5. On the other hand, the Q factor for an accumulation mode MOS varactor can be an order of magnitude higher [Porret et al., 2000]. So the Q factor for the LC tank is primarily determined by the spiral inductor. Figure 7.30 shows the layout for our spiral inductor. It is a symmetrical inductor using metal layers 3, 4 and 5. It is different from traditional spiral inductors in several aspects, and we will discuss them later in this section.

There has been tremendous research work on high-Q on-chip inductor designs [Crols and Steyaert, 1997, Lee, 1998, Mohan et al., 1999, Niknejad, 2000]. An on-chip spiral inductor can be modelled by the network in Figure 7.31 [Yue et al., 1996]. Besides the inductance L, it also has a series resistance R_s and a self-overlapping capacitance C_s. There is also the distributed parasitic capaci-

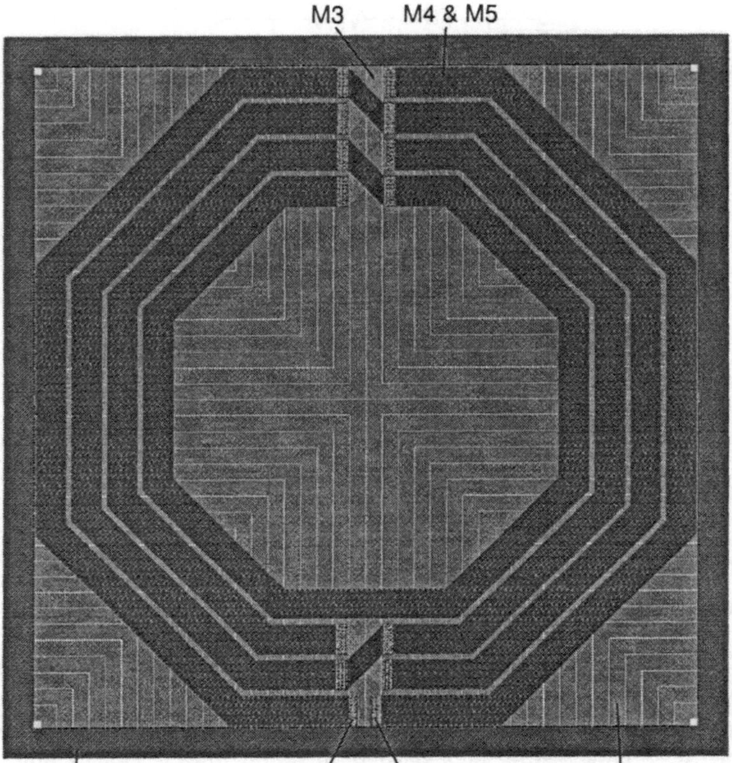

Figure 7.30. Layout for a symmetrical spiral inductor

tance to the silicon surface which is lumped into the two capacitors C_{ox} at the two terminals. Finally the high resistance substrate is modelled as a resistor R_{si} and a capacitor C_{si} in parallel.

In Figure 7.31, all the capacitors reduce the self-resonant frequency of the spiral inductor, beyond which the device becomes capacitive instead of inductive. In other words, they define the highest frequency that the device can be used as an inductor. In the same figure, all the resistors reduce the Q factor by increasing the energy loss.

There are four mechanisms that generate resistive loss in a spiral inductor:

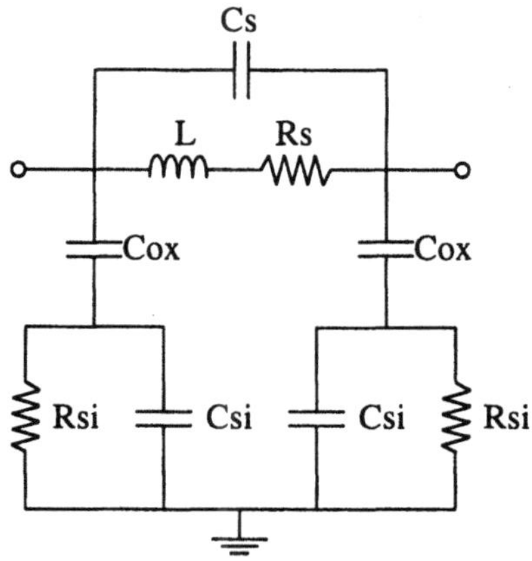

Figure 7.31. A model for an on-chip spiral inductor in a CMOS process

- metal resistance
- skin effect
- eddy current
- substrate induced loss

First, the non-zero resistivity of the metal contributes to the energy loss in the inductor. In order to achieve an inductance value of a few nH, a spiral inductor with many turns is usually needed. It often has a dimension larger than $100\mu m$. This results in a very long metal wire. The Q factor for an inductor is $Q = |\omega L/R_s|$, where R_s is its series resistance. The resistance of a wire is given by Equation 7.11, where σ is the conductivity for the metal, L, W and t are the length, width and the thickness of the wire. Even though the metal used in CMOS processes has very good conductivity, the total energy loss due to the resistance of the metal wire is non-negligible for a small inductance value of a few nH. The thickness of the metal wire is usually fixed for a given CMOS process and it cannot be changed by the circuit designer. Some CMOS processes[1] provide a thick top metal layer which can reduce the resistance of the inductor and improve its Q factor. Since the inductance value is a strong function of the number of turns and its dimension, the length of the wire is usually hard to reduce without changing the inductance. Therefore, making the wire wider can reduce its resistance based on Equation 7.11. However, this will reduce the wire spacing and increase the capacitance between the turns. It will also increase its capacitance to the substrate due to the larger area covered by the metal. Hence it lowers the self-resonant frequency and limits the highest usable frequency. Furthermore, as we will see in our later discussion, making the wire wider cannot reduce its resistance indefinitely due to some other effects.

$$R = \frac{L}{\sigma W t} \tag{7.11}$$

The metal resistivity by itself cannot explain the low Q factor for most spiral inductors. There is additional energy loss at high frequencies. One of the phenomena is called *skin effect*. At high frequencies, the electro-magnetic field cannot penetrate the conductor into its bulk. Its magnitude decays exponentially as it enters from the conductor surface. Therefore, the current is only conducted along the surface. This is equivalent to reducing t in Equation 7.11, hence increasing the resistance. The thickness of the thin layer of the conductor that is conducting is called *skin depth*, and it is given by Equation 7.12, where ω is the frequency of interest in rad/sec, μ_0 is permittivity in the vacuum, and σ is the conductivity of the metal. As the frequency ω increases, the layer that conducts electric current becomes thinner.

Aluminum is used as the primary material for interconnects in CMOS technologies. Its conductivity at room temperature is $\sigma = 3.65 \times 10^7 S/m$. Based on Equation 7.12, at 2.4GHz, the skin depth is $1.7\mu m$. From Equation 7.11, it can be concluded that if we combine two layers of metal and wind them in parallel, we can reduce the resistance, and we can reduce the resistance even further by combining more metal layers. However, due to the skin effect, using more metal layers no longer provides much help if the total metal thickness is greater than $1.7\mu m$. In our layout in Figure 7.30, we have used both metal layers 4 and 5. Adding another layer of metal 3 would not help much because of the skin effect, and it would reduce the self-resonant frequency due to the increased parasitic capacitance to the substrate.

$$\delta = \sqrt{\frac{2}{\omega \mu_0 \sigma}} \tag{7.12}$$

Another cause of the increase in energy loss is the eddy current [Craninckx and Steyaert, 1997]. An example case for eddy current generation is shown in Figure 7.32. What is shown in the figure is an outer turn and an inner turn of a spiral inductor. Assume there is an increasing counter-clockwise current in both turns at the moment shown. Now the wire closer to the center is located in the magnetic field generated by the other turns farther from the center. If the magnitude of the magnetic field is increasing, it generates clockwise eddy current as shown in the figure. The total effect is that there is more current flowing from the left side of the wire and less current flowing from the right side. This phenomenon is called current crowding. On the other hand, if the current in the coil is decreasing, the current crowding occurs along the outside edge of the wire. It is equivalent to less efficient usage of the wire cross-section and results in higher resistivity. We have only shown the eddy current generation

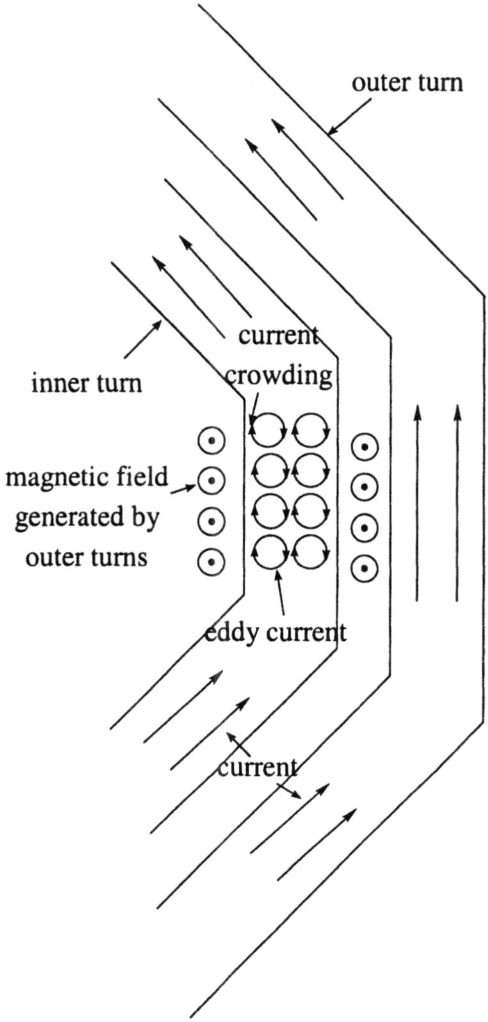

Figure 7.32. Generation of eddy current

in a small segment of the inner wire. Actually it happens on every segment of the wire. Because the magnetic field is stronger near the center of the inductor, the current crowding is more severe on the inner turns. This causes significant resistive loss at high frequencies. Therefore, in our design, we have removed the inner turns and keep the inductor hollow [Craninckx and Steyaert, 1997]. Also, because of the current crowding due to the eddy current, making the wire wider does not necessarily reduce the resistance of the inductor. The majority of the current is only conducted along the side of the wire and it swings from one side of the wire to the other during each cycle.

Finally we would like to discuss another cause of resistive loss in a spiral inductor, which is substrate induced loss. In a standard CMOS process, the spiral inductor has to stay on top of the conductive substrate, whose resistivity ranges from $1m\Omega \cdot cm$ for heavily doped silicon to $10k\Omega \cdot cm$ for lightly doped silicon. When there is current in the inductor coil, it generates horizontal eddy current in the substrate. Meanwhile, the change in the voltage potential on the wire induces vertical current in the substrate through the parasitic capacitances C_{ox} and C_{si} in Figure 7.31. The substrate current has two effects on the inductor characteristics. First, the eddy current flowing in the opposite direction to the coil current reduces the total effective inductance. Second, the resistance of the substrate causes energy loss, resulting in a reduced Q factor for the inductor.

In order to minimize the substrate induced loss, a ground shield that prevents the electro-magnetic field from penetrating to the substrate in needed to improve the Q factor and maintain the desired inductance. However, a high-conductivity ground shield induces a large eddy current in it and can greatly reduce the effective inductance. Therefore, the ground shield has to be broken into pieces to stop the path of the eddy current. A patterned ground shield was proposed in Reference [Yue and Wong, 1998]. This is also implemented in our inductor design as shown in Figure 7.30. The polysilicon layer is used as ground shield. It is broken along the path of the possible eddy current. It is also grounded via Metal 1 all around to ensure its ground potential.

Our inductor has been designed with the help of *ASITIC*, an inductor simulator software developed by Ali M. Niknejad at University of California at Berkeley [Niknejad, 2000]. In order to maintain the symmetry of our VCO, the inductor is laid out symmetrically as shown in Figure 7.30. We have chosen the radius from the center to outside edge of the outmost turn to be $R = 129\mu m$, wire width to be $W = 12\mu m$ and spacing between wires to be $s = 2\mu m$. Its inductance given by the simulation is $5.16nH$. The shape of the inductor is chosen to be octagon. In order to minimize the wire resistance, the length of the wire has to be minimized for a given inductance value. The inductance is a function of the area occupied by the coil. Therefore, the shape with the shortest perimeter for a give enclosed area is the most desirable. Mathematically the optimal shape is a circle. However, to make layout easier, we use an octagon as a good approximation to a circle.

Please also note C_{bp} in our LC VCO in Figure 7.23. Its value in our design is $10pF$. Without C_{bp} the voltage at the drain of $M1$ varies at the frequency of $2f_0$, where f_0 is the frequency of oscillator for the LC VCO. In our case, $f_0 = 2.4$GHz, and the impedance of C_{bp} is only as small as 3Ω at $2f_0$. Hence it greatly reduces the ripple at the drain of $M1$. Since the noise due to $M1$ and the bias circuit at the frequency vicinity of $2f_0$ is modulated and mixed down to low frequency, the capacitor C_{bp} reduces their noise contribution at frequencies around $2f_0$. By making the drain of $M1$ a better ground, C_{bp} also increases the frequency of oscillation slightly.

3. Simulation Results

We have simulated the PLL with a ring oscillator and the PLL with an LC oscillator with HSpice. PLLs have been one of the most difficult circuits for Spice simulations. The phase comparison process in the PFD and the high-frequency operations of the VCO and the prescaler need a small time step to achieve high accuracy. However, the phase lock capture usually takes a long time which may involve tens of thousands of clock cycles. The combination of the fine time granularity and the long capture process results in very long simulation time and many computer resources. The complexity of the circuit and the extensive involvement of the digital circuits in the frequency divider also makes the convergence of the simulation very difficult. In most cases, the accuracy of the simulation has to be sacrificed.

Despite the difficulties in simulation, we have run detailed simulations for each block and top-level simulations with relaxed accuracy to check the functionality.

For the PLL with a ring oscillator, Figure 7.33 shows the reference signal R and the signal V at the feedback divider output. Their phases are compared by the PFD. At the time shown, the PLL is still in its pull-in process and has not settled completely. Even though they look approximately in phase, their phase error can be seen from Figure 7.34 where the *UP* and *DOWN* signals are shown. In order to eliminate dead zone of the PFD, narrow pulses are generated for both *UP* and *DOWN* signals. In Figure 7.34, the VCO phase is behind the reference clock phase. Therefore, the *DOWN* signal has the minimum pulse width of about $600ps$ and the *UP* signal has wider pulse widths in order to push the VCO phase forward.

Figure 7.35 shows the simulated VCO control voltages at the outputs of the loop filter. The common-mode voltage for V_c^+ and V_c^- is kept constant by the common-mode feedback circuit of the opamp in the loop filter. The differential voltage $V_c^+ - V_c^-$ changes as a result of the wider pulse widths for the *UP* signal. As can be observed from the bottom plot, there is an upward trend for the differential control signal to minimize the VCO phase error. Meanwhile the magnitude of the ripple reduces as a result of the decreasing phase error.

Figure 7.36 shows a close-in look at the differential control signal with respect to the *UP* and *DOWN* signals. At the instance when the *UP* signal turns high, there is a current impulse sent into the active filter. Please refer to Figure 7.7. Because the opamp has a finite bandwidth and cannot respond instantaneously, it reacts as an open circuit. Then R, C_1, C_2 and the load capacitance at V_c^+ and V_c^- form a voltage divider. A pulse for the *UP* signal would cause V_c^+ to go lower and V_c^- to go higher. This initial response can be observed in Figure 7.36.

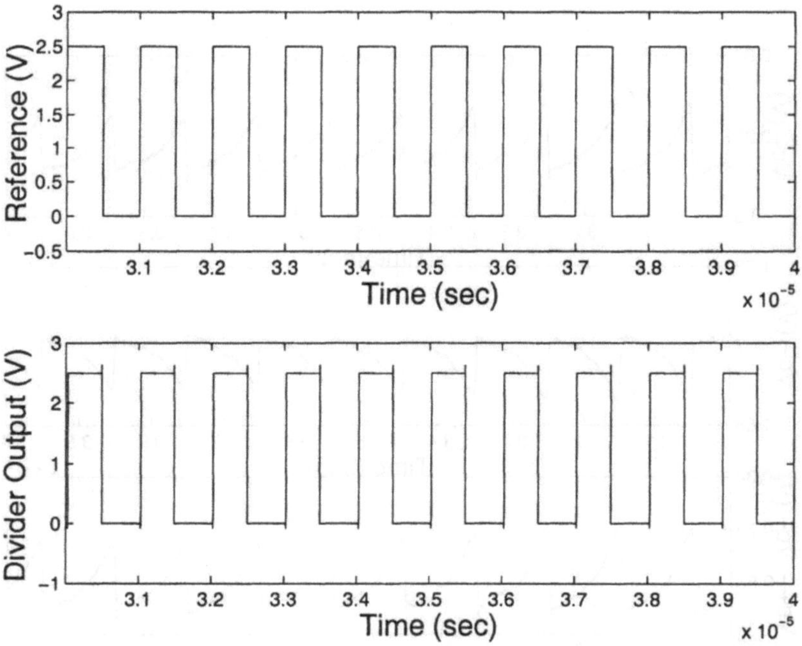

Figure 7.33. Simulated signals at PFD inputs: reference and feedback divider output

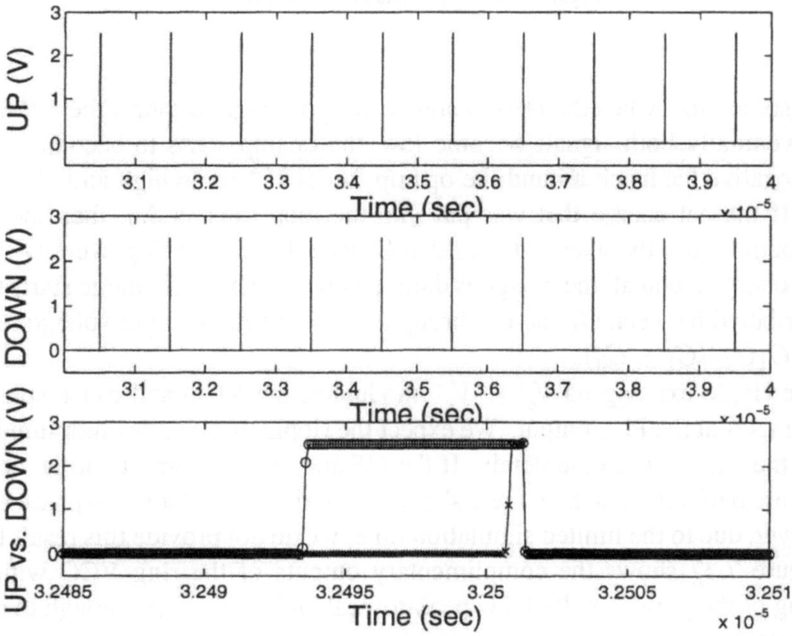

Figure 7.34. Simulated *UP* and *DOWN* voltages

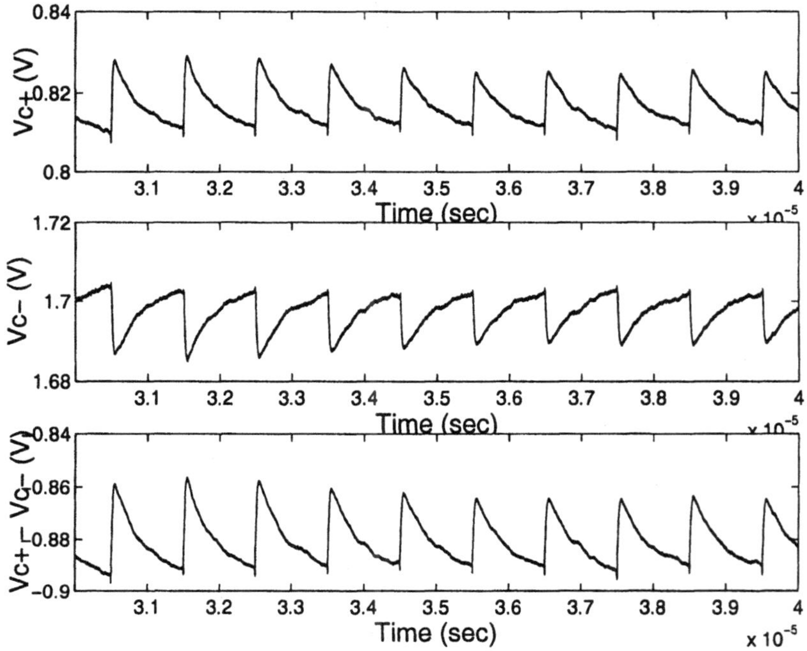

Figure 7.35. Simulated control voltages

After some time, when the *DOWN* signal also goes high to cancel the *UP* signal and eventually both signals become low, the opamp starts to become active. The negative feedback around the opamp forces V_c^+ to go high and V_c^- to go low. If the net charge that was put into the loop filter is Δq, then the loop filter output initially takes a step to $\Delta q C_2$ since the resistor R prevent C_1 from being charged, and all the charge is dumped on C_2. Then the charge starts to be redistributed between C_1 and C_2 through R, and the final output voltage settles to $\Delta q C_1 C_2 / (C_1 + C_2)$.

The ripple existing on $V_c^+ - V_c^-$ modulates the VCO and cause spurious frequencies at the PLL output. We expect the ripple to reach its minimum level when the PLL settles completely. If the *UP* and *DOWN* currents of the charge pump are perfectly matched, there should be no ripple at all for zero phase error. However, due to the limited simulation time, we do not provide this result here.

Figure 7.37 shows the complimentary outputs of the ring VCO with the loading of the prescaler. Its $1.7V$ peak-to-peak voltage is large enough to drive the prescaler stage.

Even though the above simulation results cannot be used to validate the performance of the PLL such as the phase noise, they can confirm that the

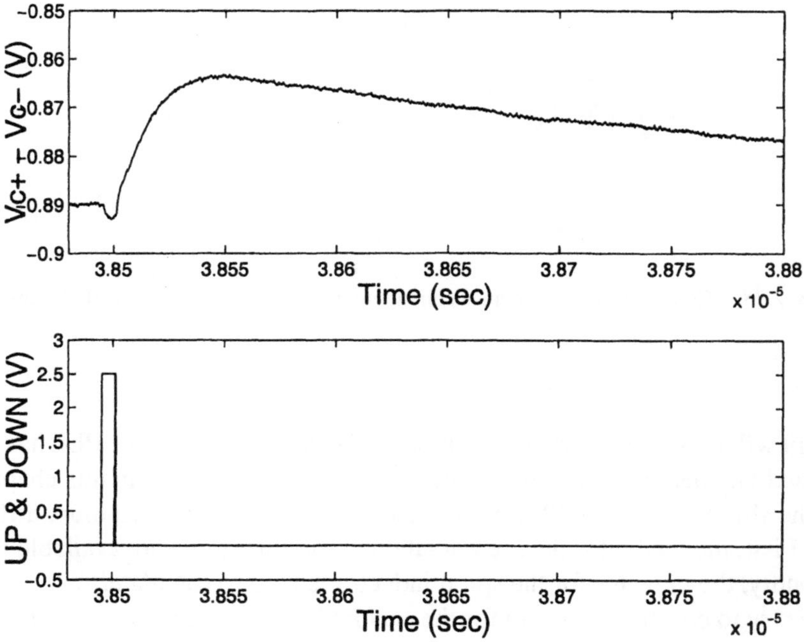

Figure 7.36. Simulated control voltages with respect to *UP* and *DOWN* signals

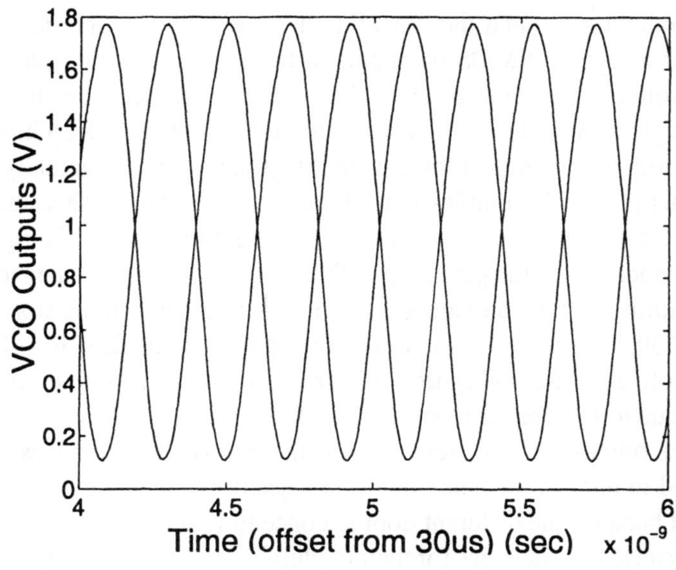

Figure 7.37. Simulated complimentary outputs of the ring VCO

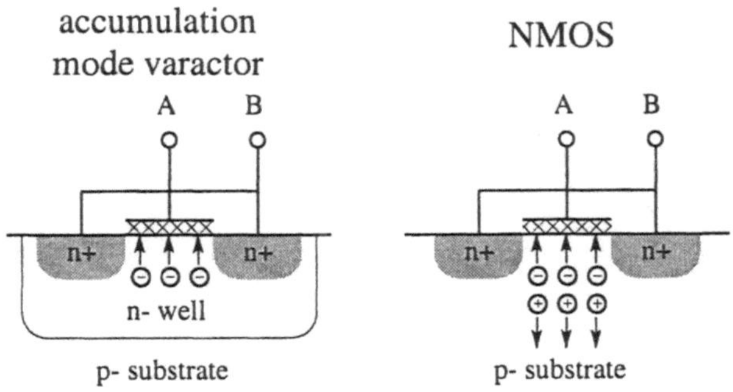

Figure 7.38. Comparison between an accumulation mode varactor and an NMOS transistor

circuit will lock under normal conditions. The performance of the PLL has been verified via measurement results which will be provided later in this chapter.

The simulation of the PLL with an LC VCO encounters even more difficulties. First, the model for the accumulation mode varactor is not available to us. Secondly, the Q factor for the spiral inductor is hard to calculate.

We try to construct a model for the accumulation mode varactors to the best of our knowledge. There is some similarities between an accumulation mode varactor and a regular NMOS in the p-type substrate working in the inversion mode. Their cross-sections are shown in Figure 7.38. When a positive voltage is applied to the gate of an accumulation mode varactor, more electrons are attracted to the surface and the N-well under the gate goes into accumulation region. For a regular NMOS transistor in the p-type substrate , a more positive gate voltage is needed to drive the p-type substrate under the gate into inversion region. As holes with positive charges are pushed deep into the substrate and electrons are pulled towards to the gate, the surface actually becomes n-type. This process resembles that of an accumulation mode varactor. The maximum capacitances for both devices are the same, which is equal to the gate capacitance due to the gate oxide. Therefore, we can create a macro model for the accumulation mode varactors by using an NMOS transistor as shown in Figure 7.39. The major difference between the two devices is a shift in the threshold voltage. Also, in Figure 7.39, a resistor R_g is added to model the gate resistance due to the poly silicon.

There are other discrepancies between the two types of devices which are not included in our model. The minimum capacitances for the two devices can be different because the different doping concentrations for the p-substrate and the N-well cause the thickness for their depletion regions to be different. Due to the same reason, their capacitance-to-voltage sensitivities can also be different. Unfortunately, we do not have enough information for the device to include

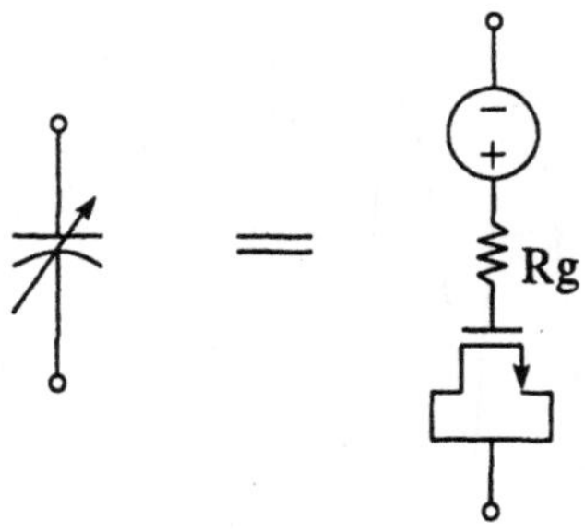

Figure 7.39. Simulated complimentary outputs of the LC VCO

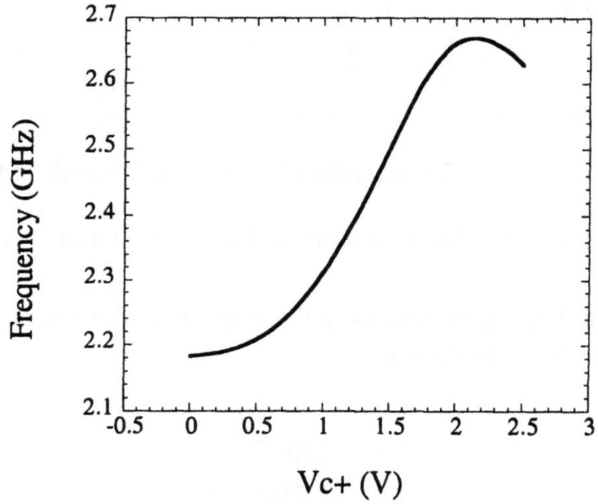

Figure 7.40. Simulated frequency characteristic for the LC VCO

them in our simple model. Since both the substrate and the N-well are lightly doped, we expect the discrepancies to be minimal.

Our calculation of the inductance value, the Q factor and all the parasitic capacitances associated with the spiral inductor is based on *ASITIC* [Niknejad, 2000]. By putting our calculated results into SPICE simulation, we have obtained an approximated frequency characteristic for the LC VCO as shown in Figure 7.40. The single-ended VCO gain can be obtained from the figure, which is $K_{VCO} = 2.3 \times 10^9 rad/(sec \cdot V)$. The droop at the top of the curve is not real for an accumulation mode varactor and is an artifact of our simplified model. Our model for the varactor includes a regular NMOS device with p-substrate. As V_c+, which is connected to its substrate, goes high, the device enters accumulation mode which does not resemble the behavior of an accumulation mode varactor with n-type substrate. Therefore, the top portion of the curve should be discarded.

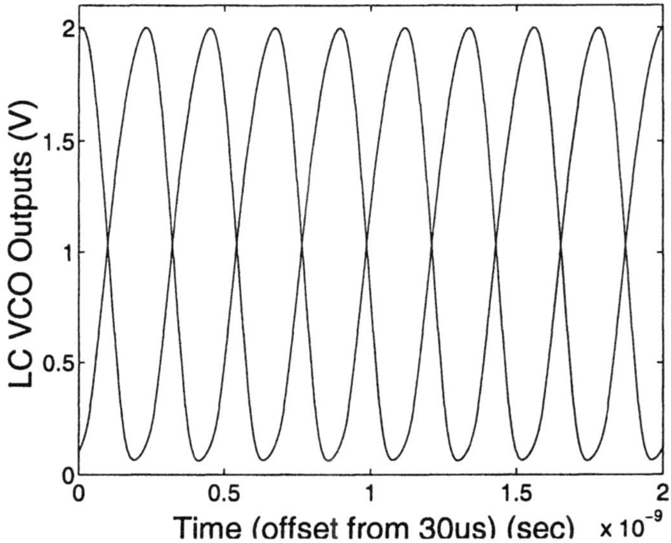

Figure 7.41. Simulated complimentary outputs of the LC VCO

Based on the VCO gain, we chose the loop filter component values and the charge pump current as follows.

$$C_1 = 139.37pF \tag{7.13}$$
$$C_2 = 8.7108pF \tag{7.14}$$
$$R = 45.677k\Omega \tag{7.15}$$
$$I_{CP} = 91.7036\mu A \tag{7.16}$$

Compared with the PLL with the ring VCO, now the VCO gain is smaller. So the charge pump current is scaled up accordingly to keep the same loop bandwidth at 100kHz. According to Equations 2.24, 2.25, 2.26 and 2.27, the in-band phase noise due to the PFD and the loop filter is expected to be lower due to the reduced K_{VCO}, while the in-band phase noise due to the feedback divider is expected to remain unchanged.

Figure 7.41 shows the simulated outputs of the LC VCO. The voltage swings at its outputs are large enough to drive the prescaler inputs.

4. Measurement Results

The PLL with a ring oscillator presented in this chapter has been fabricated in the TSMC 0.25μm CMOS technology. A micro photo of the test chip is shown in Figure 7.42 where the PLL is indicated by the dashed border, and the rest of the chip is taken by circuits for other test purposes. The PLL circuit takes a

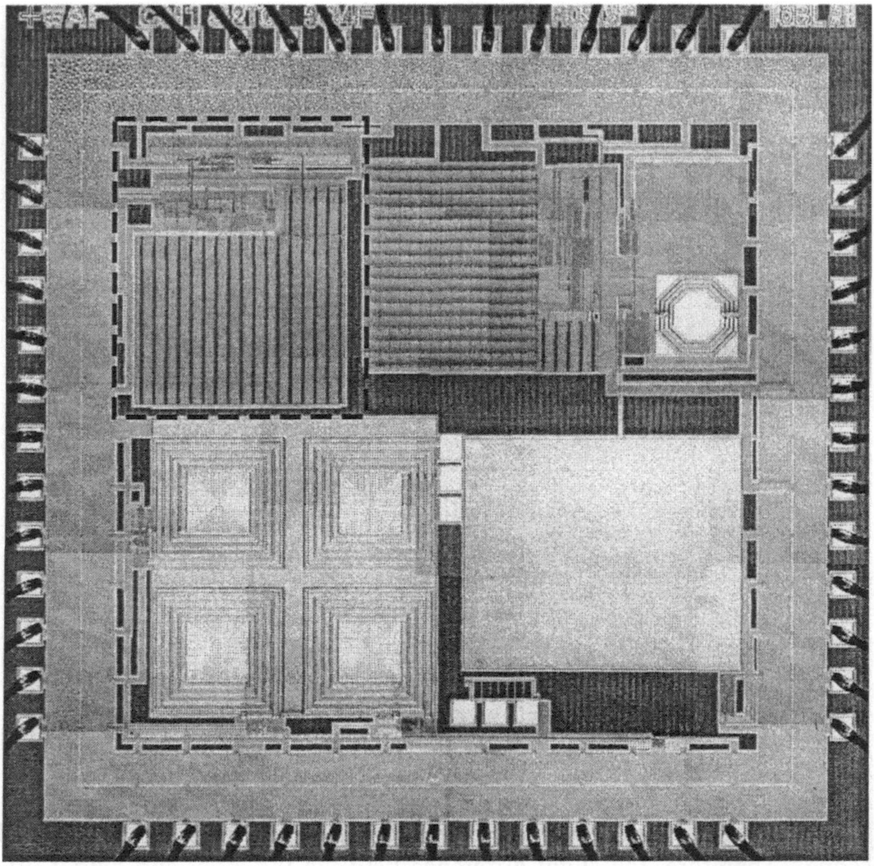

Figure 7.42. A micro photo of the PLL test chip

chip area of about $0.8mm \times 0.9mm$ excluding the bond pads, and the majority of the chip is occupied by the capacitors for the loop filter. Some measurement results will be presented in this section.

The chip is tested with a 2.5V power supply. At room temperature, the PLL locks to an input signal of up to 850kHz. The output is measured with a LeCroy WavePro 960 oscilloscope when the PLL is locked to an 800kHz input clock from an HP33120A function generator, which has been verified to be a clean clock source. The measurement is done at the feedback divider output, which has the same frequency as the reference clock. The plot for the self-referred jitter is shown in Figure 7.43 where the first line under the plot indicates the standard deviation of the period is 60ps. As a comparison, the input reference clock jitter is also measured with the same setup. The plot in Figure 7.44 shows that the standard variation of the input clock period is 100ps.

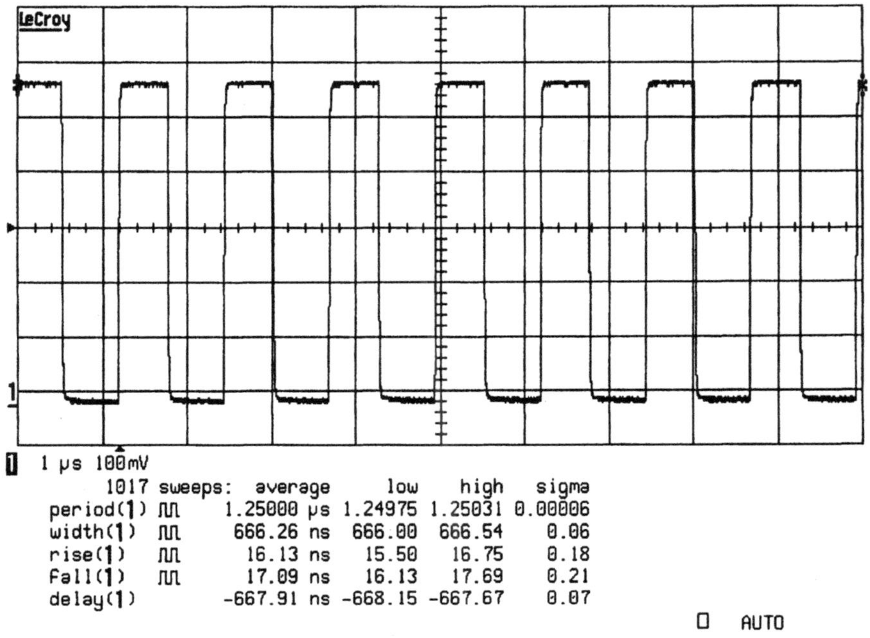

Figure 7.43. Self-referred jitter of the output clock

The accuracy of our jitter measurement is close to the limit of the oscilloscope. Therefore, the slight difference between the input and output jitter is not sufficient to show that our VCO is better than the commercial equipment. However, it does indicate that the jitter performance of our VCO is at least comparable to, if not better than, that of best commercial product on the market.

We have also captured the VCO output waveform at 1.92GHz, and the plot is shown in Figure 7.45. The RMS value for the period jitter is 3ps.

The spectrum is also measured at the feedback divider output node with an Agilent E4440A spectrum analyzer. Figure 7.46 shows the spectrum within $\pm$100kHz from the carrier, which is our loop bandwidth. It has a slope of about 6dB/oct or 20dB/dec, which is equivalent to $1/f^2$ as we have discussed before. As our reference clock is taken from a clean signal generator, the phase noise for the reference should be negligible. Therefore, it is reasonable to assume that the majority of the close-in noise is contributed by the thermal noise of charge pump and the opamp in the loop filter. The flicker noise of the charge pump and the opamp does not make a significant contribution to the total output phase noise in the frequency range of our measurement, since the $1/f^3$ phase noise region is not obvious in our measurement.

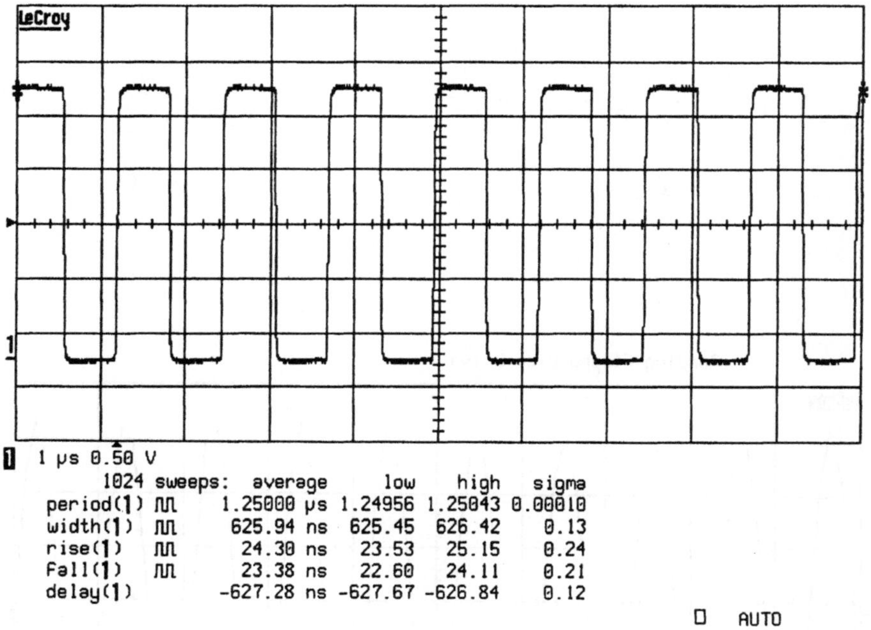

Figure 7.44. Self-referred jitter of the reference clock

Traditionally, a PLL with an LC VCO is often used for high-performance applications due to the phase noise requirement. We have demonstrated in this chapter that a PLL with a ring VCO can also serve this purpose. It can meet the phase noise requirement while saving chip area and allowing larger frequency tuning range.

5. Conclusions

In this chapter we have presented two PLL design examples. One of them uses a ring VCO and the other uses an LC VCO. We have demonstrated the design flow and considerations for the two PLLs. The circuit parameters have been optimized to reduce phase noise and ensure high-frequency operations. Also fully differential topologies have been emphasized to minimized supply and substrate noise. The test chip for the PLL with a ring oscillator is fabricated in a $0.25\mu m$ CMOS process and some measurement results for the jitter and spectrum are presented. It suggests that a PLL with a ring oscillator could be a promising candidate for high-performance frequency synthesis.

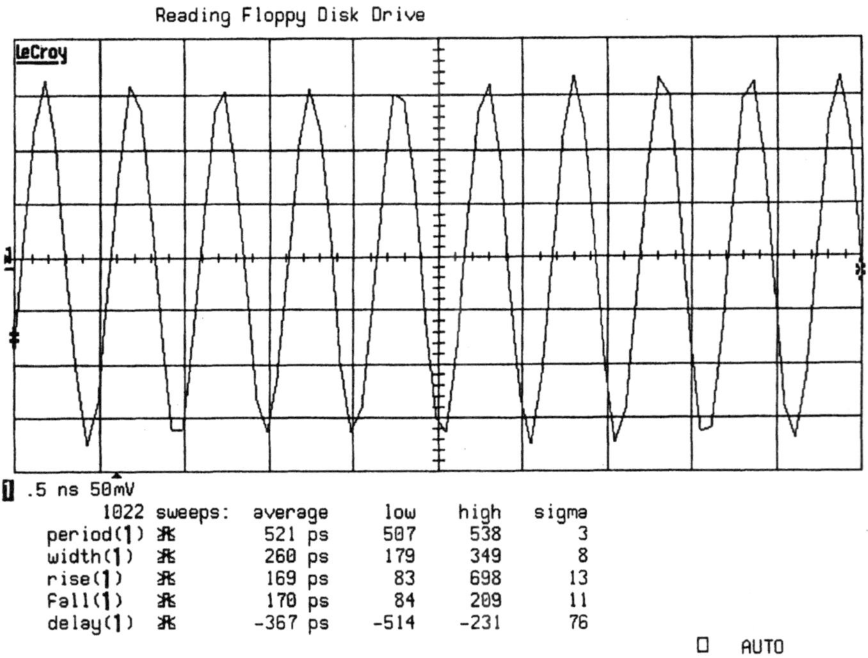

Figure 7.45. VCO output waveform at 1.92GHz

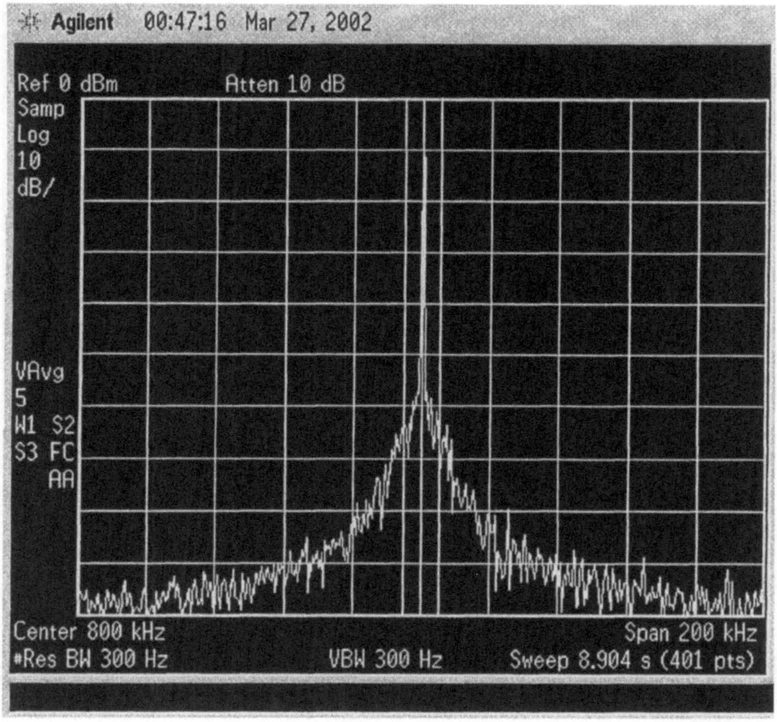

Figure 7.46. Spectrum at the feedback divider output

Notes

1 The information about some of the CMOS processes can be found at MOSIS web site at http://www.mosis.org.

Chapter 8

CONCLUSIONS

1. Research Contributions

In this book we develop a framework for CMOS VCO phase noise analysis. We attempt to bring theoretical analysis and practical designs together, and view phase noise from an angle which is different from previous work. There has been extensive research work in this area in recent years. Some of the existing work uses too many assumptions or approximations. Therefore, they become invalid in many cases. Others are mostly based on sophisticated simulations and do not provide any directions for low-noise designs. Our research work has tried to find a compromise which provides insight into low-noise designs without sacrificing the accuracy of the model.

Our research contributions are summarized as follows.

- We have developed a universal phase noise model for ring oscillators. It provides intuitive directions for low-phase-noise designs without sacrificing its accuracy. It includes the following aspects:
  - It includes a modified linear model for the additive noise contributors in ring oscillators. It suggests that fast switching is the most effective way of minimizing the additive noise.
  - We have extended the definition of Q factor to ring oscillators with voltage clipping. It predicts the improvement of ring oscillator phase noise with future CMOS technologies.
  - It includes a model for the up-converted noise in ring oscillators. It is shown that the primary noise source is in the bias circuits.
  - It includes a model for the impact of substrate and supply noise for ring oscillators.

- We have also analyzed phase noise for ring oscillators with memory delay cells. The noise sensitivity can drop by 10dB when the optimal amount of hysteresis is added to the delay cells for a 3-stage ring oscillator.

- We have designed a 2-stage CMOS ring oscillator with memory delay cells. Its phase noise performance is comparable to the measured results for CMOS LC oscillators.

- We have designed a PLL with the 2-stage CMOS ring oscillator. Its feasibility as a high-performance frequency synthesizer is demonstrated.

2. Summary

We started the book by giving a brief introduction to the application background of PLLs. In particular, we discussed the frequency synthesis in wireless communications and clock recovery in data communications which require high-speed and low-noise PLLs. Meanwhile, the demand of high-level integration with the digital circuits makes CMOS a desirable technology for implementations.

We briefly reviewed some of the basic knowledge for PLLs. We introduced commonly used circuits for each of the building blocks in a PLL and analyzed PLLs by using a linear model. More importantly, we derived the noise transfer functions from various parts in a PLL to its output. This helps us optimize loop parameters in order to minimize the output phase noise.

We also defined phase noise and timing jitter and discussed the relationship between them. Based on the way in which the measurement is made, jitter can be categorized into self-referred jitter and absolute jitter. They have different characteristics. Given phase noise expressions, we can calculate jitter, and vice versa.

Then we reviewed some of the existing phase noise models, including the ones proposed by D. B. Leeson, B. Razavi and A. Hajimiri, etc. independently. These models bring different levels of understanding for oscillator phase noise. Leeson's model is the most well-known phase noise model for LC-tank oscillators. It was proposed without any formal proof. However, it is intuitively true. Razavi's work extends the phase noise analysis to include ring oscillators by defining Q factor for ring oscillators. It is also based on a linear assumption of the circuit. Hajimiri's theory is a more complete and detailed analysis. It introduces ISF to consider the time-variant nature of ring oscillators can provide more accurate prediction for phase noise.

All the existing phase noise models have some limitations. They are either based on linear approximations or rely on massive computations. Ring oscillators are highly nonlinear when they have large signal swings, making the linear models invalid. On the other hand, the simulation-based models are usually not intuitive enough to provide help for low-noise designs.

Therefore, we have proposed a modified model which considers voltage clipping in ring oscillators. It also includes the up-conversion of the low-frequency noise. Our model focuses on the noise generation and propagation mechanism inside a ring oscillator and tries to provide intuitive design directions for low phase noise without involving extensive computations. A new Q factor for ring oscillators with voltage clipping is defined. Our study shows that the effective Q factor can be increased with faster CMOS technologies in the future. The impact of supply and substrate noise has also been studied by using ISF.

We then introduced a two-stage ring oscillator design with differential frequency tuning and quadrature outputs. It minimizes phase noise with the help of positive feedback. The differential frequency control enables fully differential charge pump and loop filters in a PLL, minimizing the impact of supply and substrate noise.

The delay cells in our ring oscillator can be modelled as delay cells with hysteresis. Our theoretical analysis indicates that the proper amount of hysteresis introduced into the delay cells can reduce phase noise. We showed some example plots for two-stage and three-stage ring oscillators with hysteresis.

We finally provided some design examples for PLLs. They are featured with differential charge pump, loop filter and VCO frequency tuning. The loop parameters have been optimized to reduce phase noise.

The theoretical work in this book is supported by extensive simulation and measurement results. We have many test chips using various CMOS technologies. They provide us the inspiration for many of our theoretical concepts and ideas.

References

[Abidi, 1986] Abidi, A. (1986). High-frequency noise measurements on FET's with small dimensions. *IEEE Trans. Electron Devices*, ED-33:1801–1805.

[Abidi and Meyer, 1983] Abidi, A. A. and Meyer, R. G. (1983). Noise in relaxation oscillators. *IEEE Journal of Solid-State Circuits*, SC-18(6):794–802.

[Craninckx and Steyaert, 1995] Craninckx, J. and Steyaert, M. S. J. (1995). A 1.8-GHz CMOS low-phase-noise voltage-controlled oscillator with prescaler. *IEEE Journal of Solid-State Circuits*, 30(12):1474–1482.

[Craninckx and Steyaert, 1997] Craninckx, J. and Steyaert, M. S. J. (1997). A 1.8-GHz low-phase-noise CMOS VCO using optimized hollow spiral inductors. *IEEE Journal of Solid-State Circuits*, 32(5):736–744.

[Crols and Steyaert, 1997] Crols, J. and Steyaert, M. (1997). *CMOS Wireless Transceiver Design*. Kluwer Academic Publishers.

[Dai and Harjani, 2000a] Dai, L. and Harjani, R. (2000a). Analysis and design of low-phase-noise ring oscillators. In *Proc. ACM/IEEE ISLPED*, pages 289–294.

[Dai and Harjani, 2000b] Dai, L. and Harjani, R. (2000b). CMOS switched-opamp based sample-and-hold circuit. *IEEE Journal of Solid-State Circuits*, 35(1):109–113.

[de Bellescize, 1932] de Bellescize, H. (1932). La réception synchrone. *L'Onde Électrique*, 11:230–240.

[Demir et al., 2000] Demir, A., Mehrotra, A., and Roychowdhury, J. (2000). Phase noise in oscillators: a unifying theory and numerical methods for characterization. *IEEE Transactions on Circuits and Systems I: Fundamental Theory and Applications*, 47(5):655–674.

[Egan, 1990] Egan, W. F. (1990). Modeling phase noise in frequency dividers. *IEEE Trans. Ultrasonics, Ferroelectrics and Frequency Control*, 37(4):307–315.

[Foty, 1996] Foty, D. P. (1996). *MOSFET Modeling With SPICE: Principles and Practice*. Prentice Hall.

[Gierkink et al., 1999] Gierkink, S. L., Klumperink, E. A. M., van der Wel, A. P., Hoogzaad, G., van Tuijl, E. A. J. M., and Nauta, B. (1999). Intrinsic 1/f device noise reduction and its effect on phase noise in CMOS ring oscillators. *IEEE Journal of Solid-State Circuits*, 34(7):1022–1025.

[Hajimiri and Lee, 1998] Hajimiri, A. and Lee, T. H. (1998). A general theory of phase noise in electrical oscillators. *IEEE Journal of Solid-State Circuits*, 33(2):179–194.

[Hajimiri et al., 1999] Hajimiri, A., Limotyrakis, S., and Lee, T. H. (1999). Jitter and phase noise in ring oscillators. *IEEE Journal of Solid-State Circuits*, 34(6):790–804.

[Joardar, 1994] Joardar, K. (1994). A simple approach to modeling cross-talk in integrated circuits. *IEEE Journal of Solid-State Circuits*, 29(10):1212–1219.

[Kinget, 1998] Kinget, P. (1998). A fully integrated 2.7V 0.35μm CMOS VCO for 5GHz wireless applications. In *IEEE ISSCC Digest of Technical Papers*, pages 226–227, San Francisco.

[Klein, 1998] Klein, P. (1998). An analytical thermal noise model of deep-submicron MOSFET's for circuit simulation with emphasis on the BSIM3v3 SPICE model. *Proc. Eur. Solid-State Dev. Res. Conf.*, pages 460–463.

[Lee, 1998] Lee, T. H. (1998). *The Design of CMOS Radio-Frequency Integrated Circuits*. Cambridge University Press.

[Leeson, 1966] Leeson, D. B. (1966). A simple model of feedback oscillator noise spectrum. In *Proc. of IEEE*, pages 329–330.

[Li et al., 2000] Li, L., Tee, L., and Gray, P. R. (2000). A 1.4 GHz differential low-noise CMOS frequency synthesizer using a wideband PLL architecture. In *IEEE ISSCC Digest of Technical Papers*, pages 204–205, 458.

[Maneatis, 1996] Maneatis, J. G. (1996). Low-jitter process-independent DLL and PLL based on self-biased techniques. *IEEE Journal of Solid-State Circuits*, 31(11):1723–1732.

[McNeill, 1997] McNeill, J. A. (1997). Jitter in ring oscillators. *IEEE Journal of Solid-State Circuits*, 32(6):870–879.

[Meta-Software, 1996] Meta-Software (1996). *HSPICE User's Manual: Software for IC Design*, version 96.1 edition.

[Mohan et al., 1999] Mohan, S. S., del Mar Hershenson, M., Boyd, S. P., and Lee, T. H. (1999). Simple accurate expressions for planar spiral inductances. *IEEE Journal of Solid-State Circuits*, 34(10):1419–1424.

[MOSIS,] MOSIS. web site at http://www.mosis.org.

[Niknejad, 2000] Niknejad, A. M. (2000). *Analysis, Simulation, and Applications of Passive Devices on Conductive Substrates*. PhD thesis, University of California at Berkeley.

[Porret et al., 2000] Porret, A.-S., Melly, T., Enz, C. C., and Vittoz, E. A. (2000). Design of high-Q varactors for low-power wireless applications using a standard CMOS process. *IEEE Journal of Solid-State Circuits*, 35(3):337–347.

[Razavi, 1996] Razavi, B. (1996). A study of phase noise in CMOS oscillators. *IEEE Journal of Solid-State Circuits*, 31(3):331–343.

[Razavi, 1997] Razavi, B. (1997). A 2-GHz 1.6-mW phase-locked loop. *IEEE Journal of Solid-State Circuits*, 32(5):730–735.

[Razavi, 1998] Razavi, B. (1998). *RF Microelectronics*. Prentice-Hall, Inc.

[Soorapanth et al., 1998] Soorapanth, T., Yue, C. P., Shaeffer, D. K., Lee, T. H., and Wong, S. S. (1998). Analysis and optimization of accumulation-mode varactor for RF ICs. In *IEEE Symposium on VLSI Circuits Digest of Technical Papers*, pages 32–33, New York, NY.

[Thamsirianunt and Kwasniewski, 1997] Thamsirianunt, M. and Kwasniewski, T. A. (1997). CMOS VCO's for pll frequency synthesis in GHz digital mobile radio communications. *IEEE Journal of Solid-State Circuits*, 32(10):1511–1524.

[van der Ziel, 1986] van der Ziel, A. (1986). *Noise in Solid-State Devices and Circuits*. Wiley, New York.

[Wang, 2000] Wang, H. (2000). A 1.8V 3mW 16.8GHz frequency divider in $0.25\mu m$ CMOS. In *IEEE ISSCC Digest of Technical Papers*, pages 196–197, San Francisco.

[Weaver, 1956] Weaver, D. K. (1956). A third method of generation and detection of single-sideband signals. *Proc. IRE*, 44:1703–1705.

[Wilson et al., 2000] Wilson, W. B., Moon, U.-K., Lakshmikumar, K. R., and Dai, L. (2000). A CMOS self-calibrating frequency synthesizer. *IEEE Journal of Solid-State Circuits*, 35(10):1437–1444.

[Wolaver, 1991] Wolaver, D. H. (1991). *Phase-Locked Loop Circuit Design*. Prentice Hall, Inc.

[Xu et al., 2001] Xu, M., Su, D. K., Shaeffer, D. K., Lee, T. H., and Wooley, B. A. (2001). Measuring and modeling the effects of substrate noise on the LNA for a CMOS GPS receiver. *IEEE Journal of Solid-State Circuits*, 36(3):473–484.

[Yue et al., 1996] Yue, C. P., Ryu, C., Lau, J., Lee, T. H., and Wong, S. S. (1996). A physical model for planar spiral inductors on silicon. In *IEEE International Electron Devices Meeting Technical Digest*, pages 155–158.

[Yue and Wong, 1998] Yue, C. P. and Wong, S. S. (1998). On-chip spiral inductors with patterned ground shields for Si-based RF IC's. *IEEE Journal of Solid-State Circuits*, 33(5):743–752.

Index

absolute jitter, 30
AMI, 74
amplitude limiter, 44
analog PLL, 10
ASITIC, 135, 141
auto-correlation function, 35

Barkhausen's criteria, 39
bias noise, 75
bit error rate, 3, 30
    BER, 3, 30
Bluetooth, 2, 107
by-pass capacitor, 80

channel charge injection, 114
charge pump, 4, 109
charge pump PLL, 13
clock feedthrough, 114, 115
clock recovery, 2
clock/data recovery, 3
    CDR, 3
closed-loop, 19, 22, 112
common mode, 89, 101
common-mode feedback, 115
    CMFB, 115
common-mode rejection, 117
conductivity, 132
coupled ring oscillator, 56, 61, 69
cross talk, 3, 28
current crowding, 133
current-starved ring oscillator, 56, 61
cyclostationary noise, 40

damping factor, 20
DC control characteristic, 101
dead zone, 15, 116
die photo, 101
die photograph, 62
differential, 101
differential control, 87, 89
differential frequency tuning, 126
digital PLL, 12
dynamic CMOS logic, 124

eddy current, 132, 133
effective Q factor (Q_{eff}), 55, 99
excess noise factor, 40, 45

f_T, 72
fast switching, 106, 149
first-crossing theory, 59
flicker noise, 29, 52, 61, 65, 71, 72, 75, 77, 80, 117
folded cascode, 116
frequency divider, 4, 9, 120
frequency modulation, 81, 88, 102
    FM, 81, 88, 102
frequency synthesizer, 1
fully differential, 56, 112

Hajimiri's model, 49
hard clipping, 68
hard limiter, 27
heterodyne, 4
homodyne, 4
HP, 69, 74, 105, 143
HSpice, 136
hysteresis, 92

impulse sensitivity function, 49, 64
    Γ, 49, 64
    ISF, 49, 64
input-referred jitter, 30

KCL, 93

LabVIEW, 105
latch, 89
LC VCO, 126

lead-lag compensation, 21
LeCroy, 143
Leeson's model, 40
linear model
    PLL, 18
linear time-invariant, 40
    LTI, 40
load pulling, 29
loaded quality factor, 40
    Q_L, 40
local oscillator, 1
    LO, 1
loop bandwidth, 22, 24–26, 107–109, 118, 142, 144
loop filter, 4, 9, 109
loop gain, 24

Maneatis load, 56, 61
Maneatis ring oscillator, 69
maximum slew rate, 69, 72, 73
memory delay cell, 150
microphotograph, 69
MiM capacitor, 111
mismatch, 102
mobility, 90
mobility degradation, 90
modified linear model, 63, 149
MOS capacitor, 110
MOS varactor
    accumulation mode, 126
MOSIS, 69, 129, 148
multiplier, 11

N-latch, 124
natural frequency, 20
noise sensitivity, 98
noise up-conversion, 74, 149

open-loop, 19, 21, 112

P-latch, 124
patterned ground shield, 135
period jitter, 31, 83
phase detector, 9
    PD, 9
phase lock, 9
phase modulation, 81, 102
    PM, 81, 102
phase noise, 3, 27
phase/frequency detector, 4, 10, 12, 119
    PFD, 4, 10, 12, 119
positive feedback, 91
power spectral density, 28
    PSD, 28
prescaler, 10, 120
program counter, 121
pull-in time, 107
pulse swallow frequency divider, 121

quaddrature, 1
quadrature, 87, 100
quality factor (Q), 47

random walk process, 30
Razavi's model, 46
relaxation oscillator, 48
ring oscillator, 5, 46, 55
RMS, 59, 64, 66, 78, 83

self-referred jitter, 30
signal-to-noise ratio, 60
    SNR, 60
single-ended, 56, 101
single-sideband phase noise, 28
    SSB phase noise, 28
skin depth, 133
skin effect, 132, 133
slew rate, 59
soft clipping, 66, 68
SONET, 2
    network element, 3
        NE, 3
    optical carrier, 2
        OC, 2
    synchronous transport signal, 2
        STS, 2
spiral inductor, 4, 46, 130
substrate induced loss, 132, 135
supply/substrate noise, 81
supply/substrate noise rejection, 88, 102
swallow counter, 121
switching noise, 88, 102
symmetrical spiral inductor, 131

tail current noise, 75
thermal noise, 52, 57, 65, 75, 76, 78, 80
thermal noise factor (γ), 57
time variant, 40
timing jitter, 3, 27
toggle flip-flop, 124
triode, 89, 109
TSMC, 74, 101, 108
two-stage, 91, 98, 108, 150

virtual ground, 115, 117
voltage clipping, 55, 149
voltage limiter, 64
voltage-controlled oscillator, 4
    VCO, 4, 9

Weaver image reject receiver, 1
white Gaussian noise, 30, 33–35
white noise, 28

MIX
Papier aus verantwortungsvollen Quellen
Paper from responsible sources
FSC® C105338

If you have any concerns about our products,
you can contact us on
ProductSafety@springernature.com

In case Publisher is established outside the EU,
the EU authorized representative is:
Springer Nature Customer Service Center GmbH
Europaplatz 3, 69115 Heidelberg, Germany

Printed by Libri Plureos GmbH
in Hamburg, Germany